The Control Method
on the Pre-release De-stress Blasting for
High-intensity Rockburst

高烈度岩爆的超前应力释放控制爆破方法

韦 猛　郑明明　陈臻林　管向丽　李 谦　著

重庆大学出版社

图书在版编目(CIP)数据

高烈度岩爆的超前应力释放控制爆破方法 = The Control Method on the Pre-release De-stress Blasting for High-intensity Rockburst：英文 / 韦猛等著.--重庆：重庆大学出版社，2020.5
ISBN 978-7-5689-2503-7

Ⅰ.①高… Ⅱ.①韦… Ⅲ.①岩爆—研究—英文 Ⅳ.①P642

中国版本图书馆 CIP 数据核字(2020)第 231531 号

The Control Method on the Pre-release De-stress Blasting for High-intensity Rockburst
高烈度岩爆的超前应力释放控制爆破方法
GAOLIEDU YANBAO DE CHAOQIAN YINGLI SHIFANG KONGZHI BAOPO FANGFA
韦 猛 郑明明 陈臻林 管向丽 李 谦 著
策划编辑：鲁 黎 范 琪
特约编辑：向 璐 邓 昊
责任编辑：范 琪　版式设计：范 琪
责任校对：王 倩　责任印制：张 策

*
重庆大学出版社出版发行
出版人：饶帮华
社址：重庆市沙坪坝区大学城西路21号
邮编：401331
电话：(023)88617190　88617185(中小学)
传真：(023)88617186　88617166
网址：http://www.cqup.com.cn
邮箱：fxk@cqup.com.cn（营销中心）
全国新华书店经销
重庆新生代彩印技术有限公司印刷
*
开本：787mm×1092mm 1/16 印张：10.25 字数：303千
2020年5月第1版　2020年5月第1次印刷
ISBN 978-7-5689-2503-7　定价：68.00元

本书如有印刷、装订等质量问题，本社负责调换
版权所有，请勿擅自翻印和用本书
制作各类出版物及配套用书，违者必究

Abstract

With the development of engineering construction in China, there are more and more tunnel projects built or planned in western China. Due to the complex geological conditions in these areas, and the deepening depth of those tunnels, the high geostress in deep tunnels usually leads to various degrees of rock bursts, resulting in personal casualties, equipment losses, delays in construction periods, or various economic risks. Many scholars at home and abroad have conducted a lot of researches on rock bursts caused by high geostress, and also put forward some preventive measures. However, since the occurrences of those rock bursts have such characteristics as uncertainty, abruptness and concealment, there still exist many disputes in the mechanism, systematic monitoring, early warning devices and preventive measures of rock bursts. In particular, there is no effective countermeasure in the engineering field to control serious high-intensity rock bursts. Thus this book studies the deformation mechanism of those high-intensity rock bursts and puts forwards a pre-release de-stress blasting method, i. e. by combining with a series of reasonable drilling and blasting parameters, an artificially controlled blasting method can be used to weaken the intensity of rock bursts, and to provide solutions to the rock bursts in high-stress tunnels and underground projects.

Based on the study of the mechanism of the pre-release de-stress blasting for those high-intensity rock bursts, combined with numerical simulations, some reasonable de-stress blasting parameters are selected, including the diameter of the borehole, the spacing between boreholes, the depth of the borehole, the angle of the borehole, and so on. Some on-site tests were also held to verify the validity of this method and the rationality of the parameters. The book has obtained the following research results:

①Through the numerical simulations of the physical and mechanical properties of surrounding rocks, and of the excavation conditions, it is found that such characteristics as compressive strength and Poisson's ratio have more obvious influence on the rock burst acitivity, while other factors as excavation size, excavation way and supporting structure have less influence. Meanwhile, through the analysis of and comparison with those prevention measures, it is confirmed that for those high-intensity rock bursts, pre-relief de-stress blasting is feasible, since it can distrub the surrounding rocks in blasting vibration, shock waves, or other huge energy, guiding redistribution of the stress in surrounding rocks, or partially releasing the stress, thereby reducing the rock burst intensity.

②Appling some related theories, such as the exponential decay theory of stess waves, we can derive a calculation formula of residual stress after de-stress blasting, which is used to find the distribution rules of the stress in the tunnel wall. The theoretical analysis and the on-site test results are in consistency.

③The numerical simulation software ANSYS/LS-DYNA and FLAC 3D were used to simulate the stress relief blasting in surrounding rocks under high-geostress conditions. According to the actual situation of de-stress blasting in high-geostress area, the numerical parameters of de-stress blasting are determined, which mainly include the diameter of the blast hole, the length of the blast hole, the spacing between the blast holes, the non-coupling coefficient of the blast hole, the order of initiation, and the included angles between the blast hole and the tunnel axis in the X direction and the Y direction, respectively. Then based on the orthogonal experiment method, an optimum blasting scheme under multi-parameter interaction conditions was determined, as well as the optimum blasting scheme under single factor variation, and the simulation analysis of de-stress blasting was also carried out.

④Based on the theoretical analysis and numerical simulation results, applied with the optimum parameters, an on-site stress relief blasting test was carried out at K186+94.020 at Sangzhuling Tunnel of Lin-la Railway. The test results show that the stress reduction rate reaches 63% (the maximum principal stress before blasting is 30.7 MPa, while the maximum principal stress after blasting is 11.5 MPa), which verifies that the de-stress blasting method has great effect on reducing the maximum principal stress. According to the rock burst grading system proposed by Tao Zhenyu, the surrounding rocks before blasting are prone to have a medium rock burst, while after blasting, the surrounding rocks have lower rockburst or even no rock burst activity. After calculating the stress in the surrounding rocks by the derived formula of residual stress after explosion, and then comparing with the on-site test results, we find the stress values in all directions are very close (with minor errors having no effect on the judging of rockbust grades), which verifies that the derived formula of residual stress after explosion in this book is scientific and reasonable.

Keywords: High-intensity rock burst; Geotress; De-stress blasting; Stress relief; Sangzhuling tunnel

Contents

Chapter 1　Introduction ··· 1
　1.1　Background and Significance　 ·· 1
　1.2　Foreign and Domestic Researches ··· 5
　　1.2.1　Current research of rock bursts　 ·· 5
　　1.2.2　Current research of geostress pre-relief controlled blasting　 ············· 10
　　1.2.3　Current research on numerical simulation of the stability of surrounding rocks······
　　　　　·· 12
　1.3　Main Research Contents and Technical Route ································· 13
　　1.3.1　Research contents ·· 13
　　1.3.2　Technical route ··· 14
　1.4　Expected Target ··· 16
　1.5　Main Innovations ··· 17
Chapter 2　Factors and Preventing Methods of Rockbursts ···················· 18
　2.1　Definition of Rockburst ··· 18
　2.2　Conventional Evaluation Indexes of Rockbursts ····························· 19
　2.3　Establishment of Numerical Model for Analyzing Factors of Rockbursts ·············· 22
　　2.3.1　Introduction to simulation tools ·· 22
　　2.3.2　Model establishment and benchmark simulation　 ······················· 23
　2.4　Impact of Rock's Physical and Mechanical Properties on Rockbursts ········· 25
　　2.4.1　Qualitative analysis of the impact of rock's physical and mechanical proportion on
　　　　　rockbursts ·· 25
　　2.4.2　Quantitative analysis of the impact of rock's physical and mechanical properties on
　　　　　rockbursts ··· 30
　2.5　Numerical Analysis of the Effects of Excavating and Supporting Conditions on Rockbursts
　　　　··· 35
　　2.5.1　Numerical analysis of the effects of full-face excavation and stepwise excavation on
　　　　　rockbursts ··· 35
　　2.5.2　Numerical analysis of the effect of supporting conditions on rockbursts ········ 38
　2.6　Recommended Ideas for Controlling High-intensity Rock Bursts ·············· 42
　2.7　Summary ··· 43

Chapter 3 Mechanism of the Pre-release De-stress Blasting for Rockbursts 45
  3.1   Definition of De-stress Blasting 45
  3.2   Microscopic Mechanism of the De-stress Blasting 46
  3.3   Derivation of the Calculation Formula of Residual Stress after De-stress Blasting ... 49
     3.3.1   Derivation of de-stress blasting equation on two-dimensional plane 49
     3.3.2   Deduction of the de-stress blasting equation on 3-dimensional plane 54
  3.4   Role of the Stress Pre-release Loose Circle 55
  3.5   Factors Influencing De-stress Blasting Effectiveness 57
  3.6   Summary 59

Chapter 4 Experimental Study on Rock Mechanics of Surrounding Rocks in Tunnels ...
  61
  4.1   Lab Environment 61
  4.2   Test Methods and Results 62
     4.2.1   Uniaxial compressive strength test 62
     4.2.2   Tensile strength test 64
     4.2.3   Normal triaxial compression test 65
  4.3   Summary 69

Chapter 5 Optimization of Parameters for Pre-release De-stress Controlled Blasting Based on Numerical Simultations 71
  5.1   Response Platforms of Numerical Simulation Analysis 71
     5.1.1   ANSYS/LS-DYNA numerical platform 71
     5.1.2   FLAC 3D numerical analysis platform 72
  5.2   Calculation Process of De-stress Controlled Blasting 73
     5.2.1   ANSYS/LS-DYNA numerical simulation blasting 73
     5.2.2   Coupling process of ANSYS/ LS-DYNA and FLAC 3D 80
     5.2.3   Stress release simulation of de-stress blasting in tunnel 83
  5.3   Scheme for the De-stress Controlled Blasting 89
     5.3.1   Overall plan 89
     5.3.2   Design for orthogonal experiment numerical simulation 91
     5.3.3   Design for single factor numerical simulation 92
  5.4   Result Analysis of Numerical Simulations 93
     5.4.1   Result analysis of orthogonal experiment numerical simulation 93
     5.4.2   Result analysis of single-factor numerical simulation 98
  5.5   Summary 103

Chapter 6 On-site Tests of Stress Release by Borehole De-stress Blasting 105
  6.1   General Engineering Situation of Test Site 105

	6.1.1 Rockburst features of Sangzhuling tunnel	106
	6.1.2 Rock bursts in Sangzhuling tunnel	109
6.2	Blasting Scheme Design	110
6.3	Test Equipment	112
	6.3.1 Stress test equipment	112
	6.3.2 De-stress blasting test equipment	113
6.4	Test Process	113
	6.4.1 Experimental process of stress testing	113
	6.4.2 Test process of de-stress blasting	115
6.5	Data Processing and Test Results	116
	6.5.1 Experimental data	116
	6.5.2 Data processing of stress test results	119
	6.5.3 Data processing results	121
6.6	Comparative Analysis	123
6.7	Summary	125

Chapter 7 Application of De-stress Controlled Blasting in Double-shield TBM Tunnel ... 126

- 7.1 Engineering Overview of Test Site ... 126
- 7.2 Influencing Factors of Rockburst Stucks in this Tunnel ... 134
- 7.3 Research Status of Advanced Rockburst Forecast Technology ... 137
- 7.4 De-stress Blasting Test Under Double Shield TBM ... 138
  - 7.4.1 Native stress test ... 140
  - 7.4.2 De-stress bursting test ... 141
  - 7.4.3 Data processing results ... 142
- 7.5 Summary ... 143

Conclusion and Outlook ... 145
Acknowledgements ... 149
References ... 150

Chapter 1 Introduction

1.1 Background and Significance

At the "2015 China (Shanghai) Technical Seminar on Underground Tunnel Engineering", according to the survey by the Tunnel and Underground Works Branch of Civil Engineering Society of China (CESC), China has become the country with the largest scale and the fastest speed in underground tunnel construction in the world. By the end of 2016, there had been nearly 13,000 road tunnels with a total length of 12,831 km in China. It is expected that there will have been a total number of 17,000 railway tunnels put into operation in China and the total length will have exceeded 20,000 km by the end of 2020. The extension of large-scale infrastructures to the west has increased not only the number of railways, highways and hydropower tunnels, but also the number of tunnels, which are longer than 10 km and demonstrate the characteristics of being "longer, larger, deeper and clustering". Some deep engineering dangers occur frequently due to the complexity of the geological conditions, including high geostress, high water head, high ground temperature and engineering disturbance. Take Sichuan-Tibet railway as an example, according to the plan, there will be 198 tunnels along the whole railway, with a total length of 1,223.451 km, which accounts for 70.2% of the total length of the line; and there are 46 extra-long tunnels with a length of 724.441 km. Among them, there will be the longest (69 km) and the second longest (59 km) tunnels in the world. In those long and large tunnels, there might appear many high geostress problems, even rock burst threats. Some relevant data show that in our country, there are an increasing number of tunnels with a depth of hundreds of meters or even over kilometers, and there also emerges some long and large tunnels crossing those deeply-buried high-geostress zones. The increase in tunnel length will inevitably lead to the rise of buried depth. Table 1.1 shows the length and buried depth of some tunnels in China.

Table 1.1 Statistics of tunnel length and buried depth in China

Name	Site	Length/m	Maximum buried depth/m
Tongyu tunnel	Kaixian, Chongqing	4,279	1,030
Qinling tunnel	Qinling, Shaanxi	(line Ⅰ) 18,460 (line Ⅱ) 18,456	1,700
The extra-long Erlangshan tunnel	Luding, Sichuan	4,176	748
Secondary tunnel of Jinping Hydropower Station	Liangshan, Sichuan	18,700	2,525
Zhegushan tunnel of National Highway 317	Aba, Sichuan	4,400	1,400
Qinling Zhongnanshan Extra Long tunnel	Qinling, Shaanxi	18,020	1,600
Ba Yu tunnel	Shannan, Tibet	13,073	2,080
Sangzhuling tunnel	Shannan, Tibet	16,449	1,347
Duoxiongla tunnel	Milin, Tibet	4,784	820

Most of these tunnels are built in complex geological conditions, and the exceeding length and corresponding buried depth make those high geostress phenomena most frequent. In recent years, furthermore, as the development of hydropower resources intensifies, the underground water power generation system has gradually developed towards the direction of super tunnel length, large house span, deep buried depth and so on, which inevitably bring more high geostress phenomena. According to some incomplete figures, by 2013 various hydraulic tunnels with total lenghth of 10,000 km had been built, over 1,000 km of headrace tunnels are under construction, and over 2,000 km of headrace tunnels have been planned. At present, some super long water diversion tunnels under construction include the Qinling Super Long Water Diversion Tunnel (98.3 km in length), the Taohe river water diversion tunnel (96.35 km in length), the Dajihuang water diversion tunnel (24.17 km in length), the Northwest Liaoning water supply tunnel (230 km in length), and the Songhuajing water supply Tunnel in Central Jilin (134.631 km in length, 6.6 m in diameter, in TBM Construction) (Kairong Hong, 2015). Moreover, there are countless large-scale, complexly structured underground cavern groups, built for various uses, crossing each other in the limited space.

Due to the reduction and even depletion of shallow resources, deep well mining has become imperative and a vital technology in mining industry throughout the world. In China, the coal reserves buried in less than 1,000 meters only account for 53% of the total coal resources. At present, the depth of coal mine is increasing at a rate of 8 to 12 m every year, and the mining depth in eastern China is even developing at a rate of 100 to 250 m every 10 years. It is even expected that many coal mines will enter the depth of 1,000 to 1,500 m in the next 20 years. In recent years, some metal mines in China have entered deep mining layers, such as the mining depth of Huize Pb-Zn Mine in Yunnan Province has exceeded 1,000 m, the Dongguashan Copper Mine in Tongling has reached 1100 m, and the Hongtoushan Copper Mine in Fushun has reached 900 - 1,100 m. Over seas, there are nearly 100 metal mines with a mining depth of over 1,000 m, among which the gold mine of AngloGold company in western Africa has reached 3,700 m; in Kolar gold mining area of India, there are 3 mines with a depth of over 2,400 m, with one even reaching 3,260 m; in Krivolog iron mining area of Russia, the mining depth is up to 1,570 m. In addition, the mining depth of some metal mines in Canada, the United States and Australia has also exceeded 1,000 m.

While China is vigorously developing the western region and promoting the economy of the western area, we need to pay special attention to the inevitable deep engineering problems, i. e., high geostress, uneven distribution, complexity and difficulty of construction procedures. In a condition of high geostress, the excavation of underground tunnel often produces severe rock bursts, which directly threatens the safety of construction personnel and equipment, and affects the progress of the project, so it is urgent to put forward a proper solution to avoid those geological disasters.

Based on incomplete statistics, at least 2,000 coal bursts or rock bursts have occurred in many coal mines in China. In some severe rock bursts, tons of rock blocks, rock slices or rock plates have been thrown out, causing huge damages. Rock bursts also occur frequently in some hydropower projects, such as in the underground works of Ertan, Taipingyi, Jinping, Tianshengqiao and other hydropower stations. According to some document retrieval, at least 18 countries or regions in the world have large-scale rock bursts. Thus the stability of the surrounding rocks in high-geostress tunnels has become an urgent problem to be solved in the excavation of underground tunnels. Numerous engineering projects show that the greater the buried depth, the more intensive the tectonic movement, and the higher the geostress level, the higher the possibility of deep engineering threats, such as high-intensity rock bursts, continuous deformation and even large-scale collapses. These dangers have seriously restricted the development of hydraulic and hydroelectric projects, transportation, national defense, and deep basic physics in China, and even affected the safety of mining in China. In order to reduce or avoid the rock bursts, and minimize the economic losses and casualties, it is urgent to conduct in-depth researches on the mechanism, forecast, and preventing techniques of those high-intensity rock bursts.

The Control Method on the Pre-release De-stress Blasting for High-intensity Rockburst

In underground engineering, the fundamental reason for the instability of the surrounding rocks is that the excavation causes the concentrated stress to exceed the strength of the surrounding rocks, which makes them break and even destroyed, thus forming a broken zone. Therefore, in order to keep the stability of the surrounding rocks, some control measures need to be taken to prevent the forming of a broken zone, or to restrain the developing of the broken zone. On the contrary, if the concentrated stress is less than the strength of surrounding rocks, they will not be damaged, causing no broken zone, and the tunnel will be in natural stability without any treatment. Specifically, two measures can be taken: one is the strengthening method, i.e., to strengthen the surrounding rocks to improve their intensity; the other is the geostress control method, i.e., to control the geostress distribution in the surrounding rocks, so that the peak geostress is transferred to the deeper rocks far away from the tunnel face, to avoid high-geostress concentration or tensile zone around the tunnel, keeping the tunnel in low-stress area. According to Xiangshen Guo (2010), the strengthening method is suitable for ordinary tunnels, while for the deep-buried, high-stress tunnels, the method is not so ideal as it may cost more labor force, more materials and more financial resources, usually with high supporting costs.

So far, for high-intensity rock bursts, scholars have made many in-depth research and lots of scientific achievements, most of which, however, just focus on forecast and evaluation. The control measures are fewer, even the existing ones, to some degree, have certain limitations, without the control of solving the problem of high-intensity rock burst. Furthermore, researchers today investigate rockbursts mainly through field investigation and laboratory test, combined with mechanism analysis of existing physical properties and mathematical methods. Take 3 methods as examples: ①The high-pressure water injection method. According to Xianneng Wang (1998), this method is to make advancing boreholes in rocks and then inject water into rock mass with high pressure and even flow. This method can release strain energy and transfer the maximum tangent stress to deeper surrounding rocks, which might generate some new tension cracks and make the original cracks expanding to weaken the intensity of rock mass, thus reducing the capacity of the rock mass to store strain energy. However, as is shown in the case of the Krishna Gold mine in Kolar India, for those hard rock mass with high geostress, water injection might bring new problems by triggering some rock bursts around the internal cracks, so the method has certain limitations. ②The staged excavation method. Many scholars recommend the staged excavation method to reduce rock bursts. However, this method, according to the numerical analysis of the Tianshengqiao II Hydropower Station, is not necessarily beneficial because the more excavations, the higher chances of rock bursts. ③The reinforcement method. In high-intensity rock burst sections, we can take such measures as deepening and densifying system anchor bolts, adding base plates, hanging a whole network, spraying and mixing cement. However, as shown in a series of engineering practices, when anchor bolts are used in areas with severe rock bursts, it is prone to cause accidents such as collapse. All in all, the above methods have no obvious effect on reducing those high-intensity and severe rock bursts.

To sum up, there is no effective solution to the high-intensity rock bursts under the high-geostress condition, so it is necessary to explore new forecast and control measures. The author of this book holds that to control blasting by pre-releasing geostress can effectively weaken, and even prevent those high-intensity rock bursts since it can reduce the stress accumulation in surrounding rocks before large-scale excavation. Therefore, it is of vital importance to study the control method on pre-relief de-stress blasting for high-intensity rockbursts. In some areas with a high proneness of rockbursts, before unloading excavation, we can release the geostress of surrounding rocks by applying the control de-stress blasting technique to reduce the energy concentration and to wipe out rock bursts from the root. This study might also provide some theoretical support for the treatment of high-intensity rock bursts in deep engineering projects.

1.2 Foreign and Domestic Researches

China is experiencing a rapid development of infrastructure construction. So the research on the evolution, forecast and prevention of deep engineering dangers is an important topic for the safety construction and operation of hydraulic and hydroelectric projects, transportation, national defense projects, as well as for the safety and efficiency of metal mining industry. All these deep engineering projects, involving the fields of mining, transportation, water conservancy and hydropower, national defense and others, are directly related to the sustainable development of national economy. Rock burst is a moving geological threat occuring in hard rock tunnels in high-geostress areas. There are usually a large number of blocks exploding in the excavation space when rock burst occurs, which often cause huge damage to personnel and equipment and affect the progress of the project. Therefore, rock burst, as a typical threat in high-geostress area, has always been a hot topic in the field of geotechnical engineering. Scholars at home and abroad have done a lot of researches in this field.

1.2.1 Current research of rock bursts

The earliest record of rock burst in the world can be traced back to the 1830s, when a rock burst occurred in the Leipzig Coal Mine in UK. Since then, rock bursts of various sizes have occurred in underground projects around the world. We made rough statistics as follows: in Norway, a strong rock burst took place in Sima underground power station; in Sweden a rock fragment ejection with sounds also took place in Forsmark nuclear power station; in South Africa, some rock bursts even triggered earthquakes in mine pillars, mining zones, fault zones; Canada also experienced two large rock bursts, which triggered fault sliding, collapses, earthquakes and casualties, and finally the mines were shut down; in Japan rock bursts also occurred in the Guanyue Tunnel with a total length of 10.9 km, of which 1.1 km was the rock burst area, where most rock bursts occurred in quartz diorite instead of hornfels, and in place without water gushing, not in place with water gushing; in Guizsasso Road Tunnel in Italy a rock burst occurred suddenly on the right side of the tunnel face, and the top arch collapsed, with a burst volume of hundreds of cubic meters.

Because of its suddenness in time, randomness in space, burst ejection in form, and harm in result, rockburst has attracted worldwide attention. In particular, South Africa, as one of the most advanced countries in rock burst research in the world, has accumulated rich experience in theoretical research, forecast, prevention and other control measures of rock bursts, and has established a fairly complete theoretical system. In China, the research history of rock burst is relatively short, with the earliest record of a rock burst in Shengli Coal Mine, Fushun, in 1933. Since the late 1970s, rock burst research has entered a new stage both at home and abroad. Many international academic conferences on rock bursts have been held for exchanging and accumulating valuable data and practical achievements on the forecast, control and prevention of rock bursts. In recent years, the main domestic research of rock bursts include the follows: Tong Jiang et al. (1998) summarized the main rock burst theories, discussed their advantages and disadvantages, and forecasted the future developing trend; Shaohui Tang (2003), Linsheng Xu et al. (2001; 2002; 2004) introduced some actual rock bursts and field researching results; Yuanhan Wang et al. (1998) put forward some methods to judge the occurrence and intensity of rock bursts; Tianbin Li et al. (2011) introduced some research results in field simulation tests; In combination with actual engineering practices, Guoqing Chen et al. (2013), Xiangdong Xu (2008), Peng Yan et al. (2008), Gong Fengqiang et al. (2007), Wang Bo et al. (2007), Guan Jianji et al. (2006), ManChao He et al. (2006), Yongmou Xie et al. (2004), Zemin Xu et al. (2004; 2003), Yun Jiang (2002), and Jianglin Wan et al. (1994) also disscussed the mechanism, forecast methods and prevention measures of rock burst from different aspects, and obtained lots of beneficial results. Through literature review, we can summarize the main achievements as follows:

(1) Classification of rock bursts   The classification of rock bursts is mainly based on the storage and release of elastic strain energy in rock mass or the patterns of geostress effects. At present, there exist some disputes among scholars. Zhuoyuan Zhang et al. (1994) divided rock bursts into three types according to the site and the amount of energy released, namely the rock burst caused by sudden rupture of the surface rock on the cave wall, the rock burst caused by

sudden failure of ore pillar or of large range of surrounding rock, and the rock burst caused by fault movement. Based on the intrinsic factors, Wenzhi Zuo et al. (1995) classified rock bursts into three types: the horizontal tectonic geostress type, the vertical pressure type and the comprehensive type. The research group of Tianshengqiao Ⅱ Hydropower Station proposed two standards for classification. Firstly, according to the degree of rupture, rock bursts can be divided into two categories: the fracture relaxation type and the rupture detachment type. Secondly, based on the scale, rock bursts can be divided into three categories: the sporadic rockbursts (0.5–10 m long), the monolithic rockbursts (10–20 m long) and the continuous rockbursts (larger than 20 m). Zhi Guo (1996), based on the failure modes, classified rock bursts into three types: the ejecting type, the flaking type and the collapsing type. Up to now, the most influential classification comes from Dr. Tan Yi'an: according to the origin of high geostress and the direction of maximum principal stress, he firstly divided rock bursts into three categories, namely the horizontal geostress type, the vertical geostress type and the mixed geostress type; secondly, according to the specific conditions of geostress and features of rock bursts, he divided rock bursts into six sub-categories.

(2) Intensity of rock bursts   Nowadays there also exist different opinions on the intensity of rock bursts. In 1981, according to the scale of damage to the project, German Bukhoino divided rock burst intensity into three levels: the minor damage, the moderate damage and the severe damage. In 1974, while studying steep slope tunnels in Norway, Norwegian Russeness B. F. defined four grades from 0 to 3, according to the sound features and the failure characteristics of surrounding rocks. In 1988, Dr. Tan Yian divided rock burst intensity into four levels: weak, medium, strong and extremely strong, according to the damage level, the mechanical and acoustic characteristics and the damage pattern. In 1996, in the *Technical Consultation Report on High Geostress in Erlang Mountain Tunnel* issued by China Railway No. 2 Research Institute, rock burst intensity was divided into three levels: minor, moderate and severe, according to the criterion of σ_θ/R_b. In 1996, based on the sound features, the deformation fracture of rocks, the σ_θ/R_B ratio and the maximum horizontal principal stress σ_{Hmax}, σ_{Hmax}/σ_V, the First Highway Design Institute of Ministry of Communications of China divided rock burst intensity into three levels: minor, moderate and severe. Based on previous classifications, the research group on High Geostress and Rock Stability of Erlangshan Tunnel on Sichuan Tibet Highway, headed by Professor Lansheng Wang, defined four levels: slight, medium, strong and severe, based on the damage scale, the sound, movement, form characteristics, the position, the aging feature, the depth and the σ_θ/R_b ratio, etc.. Based on those previous research and combined with domestic engineering experience, Jingjian Zhang and Bingjun Fu (2008) pointed out that no rock burst will occur when $\sigma_c/\sigma_1 > 14.5$ and a rock burst might occur when $\sigma_c/\sigma_1 \leq 14.5$, and they also divided rock burst

intensity into four levels.

(3) Mechanism of rock bursts At present, many scholars at home and abroad have applied various theories to analyze mechanism of rock burst from the aspects of strength, stiffness, stability, energy, fracture, damage and mutation theory, etc. They have also put forward many hypotheses, as well as formed different theoretical criteria, which have had a certain degree of agreement with some practical projects. E. Hoek et al. believed that rockburst was the result of shear failure of the surrounding rocks in high-geostress area. When explaining the cause of borehole collapse, Professor Zoback also believed that the hole wall collapse, which was similar to "rockburst", was just a kind of shear failure. However, Mastin (1984) and Haimson (1972), after conducting a unidirectional compressed physical simulation test on a sandstone slab with a circular hole and reproducing the collapse phenomenon of the hole wall in the laboratory, believed that this phenomenon was caused by the partial failure of the stress concentrated part of the hole wall and was the product of tensile fracture. Professor Shuqing Yang et al., at the diversion tunnel of Tianshengqiao II hydropower station, after making physical simulation experiments on some similar materials, summed up two mechanisms for rock fracture and shear, and pointed out that they were the products of two kinds of stress: the fracturing damage belongs to the brittle fracture, and the shear failure is the destruction in the peak strength of geostress. Yi'an Tan (1992) thought that rock burst is an asymptotic failure process, and its formation process follows three stages: splitting into plates, shearing into blocks and ejecting into pieces. "The Research Group on High Geostress and Rock Stability of Erlangshan Tunnel on Sichuan Tibet Highway", led by Professor Lansheng Wang, compared the rock burst effects with the whole process of rock deformation and failure under the condition of three-direction geostress, and concluded that the mechanical mechanism of rock burst can be summarized into three basic forms, namely, the compressive tensile crack, the compressive shear tensile crack and the bending and folding, which can appear in combined forms.

(4) Forecast of rock bursts While lots of researches have been done in the forecast of rock bursts at home and abroad with huge achievements, there still exists some limitations in research and methods, due to the complexity of forecast. The forecast methods of rock bursts in the world can be roughly divided into the theoretical analysis method and the field test method (Li Guo, 2003). The theoretical analysis method has advantages in the developing trend forecast, since it costs low and it can apply the existing data to get a rapid evaluation of the rock burst trend. The field test method is to use necessary instruments to directly monitor or test the rock mass to determine the possibility of a rock burst, or to indicate the approximate time of the rock burst, so as to pull out staff and equipment in time and to ensure safety. The field test methods includes various direct contacts and those geophysical patterns.

(5) Prevention and control of rock bursts In terms of prevention and control of rock bursts, various studies have been conducted at home and abroad. In the past the main measures were as follows: ①to reduce disturbance to surrounding rocks and concentration of geostress. In the section with a high proneness of rock bursts, a short footage driving method shall be adopted as far as possible, since it can lower the explosive amount used for primary blasting, and reduce the disturbance to the surrounding rocks; at the same time, a smooth surface blasting should be applied to smooth the excavation contour and to avoid geostress concentration caused by uneven surface. ②to spay water. To spray water to the new excavation face after blasting, so as to reduce the capacity of rock mass to store strain energy; ③to make immediate support. In order to lessen the geostress of surrounding rocks, immediate and effective support should be provided to the section where rock burst occured or may occur. ④to make advance borehole. In order to release high geostress beforehand, some advance boreholes are drilled in the tunnel face.

In conclusion, although huge achievements have been made in the study of rockburst at home and abroad, there are still some problems yet to be solved, due to the complexity of rockbursts and the diversity of geological environment. These problems include:

Both at home and abroad, the intensity classification of rock burst is usually based on one single or a few indexes, not on comprehensive factors. There are gaps between theory and its applications. For example, in the grading scheme of rock burst intensity, on the rock burst sites, insufficient attention is paid to the macroscopic signs or phenomena which are easy to identify, thus making it difficult for theoretical researchers and on-site technicians (engineering designers, construction and supervision personnel, etc.) to reach consensus in a timely manner, so their views are often neglected on the site. There are also different views about the mechanism of rock bursts, without classical and unified theory consented by all academic communities. Most rock burst theories or criteria do not consider the geostress concentration caused by excavation of caverns (the loosening and fracture of surrounding rocks), or they only take one factor into consideration, thus the results are not systematic and comprehensive, and some assumptions are even lack of theoretical and experimental evidence. The measures for controlling rock bursts, such as injecting water or drilling, do have certain effects to reduce rock burst intensity, especially for those medium and minor rock bursts. However, in high-intensity rockburst areas, due to the intactness of the surrounding rocks, the underdevelopment of cracks, the poor splitting effect of water injection, and the limited release capacity of empty holes and channels, etc., the above-mentioned methods are insufficient to meet engineering needs, so it is urgent to explore more scientific and effective methods.

1.2.2 Current research of geostress pre-relief controlled blasting

For avoiding intensive unloading after excavation, to pre-release the geostress in surrounding rocks before excavation is a better choice. Geostress pre-relief technique is adaptable to various geological conditions, especially suitable to hard rock tunnels and deep buried tunnels. In 1950, this technique was applied in a gold mine in South Africa, and successfully reduced rock bursts, by improving the geostress in rocks of hanging walls. This technique was also applied in lead-zinc mines in India. The research on geostress pre-relief technique in China started at the end of 1980s: Zhaofang Xing introduced it into the field of outburst prevention and put forward an anti-outburst measure of controlling stress-relief blasting in deep holes. Bole et al. (1993a, 1993b) analyzed many successful cases of rock burst prevention by blasting, and established a preliminary mathematical model, but the model had some limitations. Based on Livingston's energy balance theory, Shoufeng Chen et al. (2001) put forward a designing plan of stress-relief controlled blasting in surrounding rocks under high geostress. Shudong Feng (2007), Jianfeng Li et al. (2008) examined the control effect of loose blasting on percussive rock bursts, expounded on the blasting mechanism, and introduced the blasting technique and the parameters of loose stress pre-release. Based on the features of tunnel rock bursts, Jiande Cai et al. (2008) put forward a method of stress-relief blasting, and they, according to numerical simulations, also optimized the stress-relief blasting plan for extended auxiliary shallow holes.

Mazaira, Konicek, etc. found that in the excavation of deep tunnels, when the lithology of the surrounding rock is hard, it is easy to produce high stress concentration in the surrounding rocks of the excavation boundary. As shown by the solid line in Figure 1.1, the hard surrounding rock, which has high strength and complete shape, will accumulate a large amount of elastic strain energy under high geostress. In excavation, it is easy to have dynamic instability (brittle failure), so the rock burst risk is high. After the surrounding rock is loosened by blasting, certain crushing area or fracture area will be produced, where a large number of cracks will replace the original intact rock, thus degrading the mechanical force of the surrounding rock, reducing its modulus, and causing plastic failure. Meanwhile within the range of the blasting, the stress in the surrounding rock is greatly reduced, and the peak geostress is transferred to those deep elastic zones, and large amount of elastic strain energy accumulated in the surrounding rock is released, as is shown by the dotted line in Figure 1.1, and thus the risk of rock bursts is reduced.

In conclusion, some achievements have been made in the research of controlled blasting technique for de-stressing surrounding rocks, but a series of problems still need to be solved:

①The geostress pre-relief technique is now mainly applied in some small section

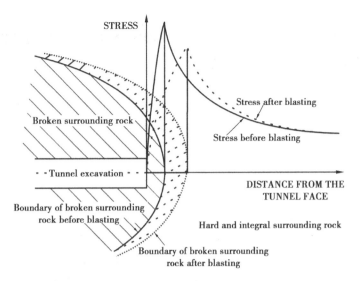

Figure 1.1 The mechanism to control rock burst using de-stress blast

tunnels. However, due to high frequency of rock bursts, it is seldom applied in those deep-buried large-section tunnels with higher geostress and more obvious unloading effect in large scale excavation.

②The interacting mechanism between the advance de-stress blasting and the geostress in surrounding rocks is not clear;

③The related parameters of de-stress controlled blasting, including diameter of charge, distance between holes, etc., need to be verified by field tests to achieve refinement and standardization.

④There is a lack of a more direct, accurate and convenient evaluation mode for the effects of de-stress blasting, owing to the difficulty in testing geostress change in deep surrounding rocks of underground projects. The quantification of geostress change is a main reference for evaluating the rationality of blasting parameters and for the feedback of the controlled pre-release de-stress blasting.

Therefore, it is of great significance to make research on the control method of pre-release de-stress blasting for high-intensity rockburst, including the research on the interacting mechanism of the on-site geosress and the controlled pre-release de-stress blasting, the research on the scope of loose zones, and the research on the criterion of blasting parameters, all of which are carried out to meet the major needs of national construction, to provide solutions to rock burst threats faced by many deep buried tunnels and underground projects.

1.2.3 Current research on numerical simulation of the stability of surrounding rocks

To study the stability of tunnel surrounding rocks, researchers mainly apply the experimental research, the theoretical research and the numerical simulation research, among which the numerical simulation research, owing to its convenience and low cost, has been widely used. This research includes the finite element method, the finite difference method, the discrete element method, and so on.

The finite element method, as a method of deep theoretical basis and wide application, can solve the problems that can not be solved by analytical methods in the past. It is also very effective to solve those complex problems with irregular boundary and various structures. Jun Xu et al. (2003) deduced the calculation iteration formula of the elastoplastic random finite element incremental initial stress method. Combining the reliability theory, they proposed a reliable analysis method to test the stability of surrounding rocks, and also pointed out the geostress values of the surrounding rocks and the anchor and shotcrete support structure. Huabing Zhang et al. (2004) used the viscoelastic plastic model to make finite element analysis of the surrounding rocks in loess tunnel, and found that the simulation results are basically consistent with the deformation and failure process in real tunnels. Wenqing Hu et al. (2004) took the construction of the weak surrounding rock section of the Muzhaling Tunnel as an example, and carried out a finite element numerical analysis of the plane elastoplastic surface. Gang Liu et al. (2003) used ANSYS software to calculate the loose zone of surrounding rocks in rectangular tunnels, and obtained the qualitative and quantitative relationship between the loose zone and its influencing factors. Dehai Li et al. (2005) used the corresponding principle of viscoelastic theory to get the theoretical solution to the displacement field of the viscoelastic model in the axisymmetric circular roadways. By way of numerical simulation analysis in ANSYS, he also made comparative analysis between the theoretical analytical solution and the numerical simulation solution.

The finite difference method is mainly used to solve large deformation problems which can not be solved by the finite element method. Based on the principle of finite difference method, FLAC (Fast Lagrangation Analysis of Continuum) numerical analysis method is proposed. Some scholars used the hoek-brown empirical criterion, based on the GSI method, and the strength reduction method of fast Lagrange difference method and others to determine the strength parameters and deformation parameters of the excavated tunnel. This method is most suitable for the discontinuity and large deformation of rocks and soil mass, and the solution speed is faster, but the disadvantages are that it is not easy to solve the boundary problem and the mesh division is more arbitrary. The FLAC 3D uses the program of three-dimensional explicit finite difference, which can simulate the mechanical behaviors of geotechnical materials. In the calculation, the Lagrange algorithm does not form a stiffness matrix and does not need to iteratively satisfy the elastoplastic constitutive relationship. Instead, it only needs to calculate geostress through strain, which greatly saves memory and time, compared with the ordinary implicit solution method. The explicit time approximation method of FLAC 3D general equation of motion is more suitable for progressive failure and instability of rocks and soil, as well as large deformation analysis. ANSYS/ LS-DYNA is a specialized blasting simulation software, which can dynamically simulate the instant mechanical changes in tunnel blasting and other issues, and has been approved by many designers and researching institutions. Therefore, in the study of de-stress blasting, we plan to use ANSYS/ LS-DYNA software to carry out dynamic simulation of controlled de-stress blasting in the hope to explore more clearly and vividly the influence of parameters on the de-stress effect, and to play a guiding role in the simulation analysis of high-intensity de-stress controlled blasting. FLAC 3D can well solve the stress problem of surrounding rocks under the static state. For the design of the tunnel stress-relief scheme, there are few assumptions about the numerical simulation analysis, so it is close to the actual deformation and boundary conditions of the surrounding rocks. Therefore, in this project, ANSYS/ LS-DYNA is used to simulate the instantaneous action of blasting, and then the FLAC 3D software is used to simulate the change process of geostress field under the static action. The technical means of combining the two software data is used to study the application effects of de-stress blasting scheme in high-geostress tunnels.

1.3 Main Research Contents and Technical Route

1.3.1 Research contents

Based on the mechanism of pre-release de-stress controlled blasting for high-intensity rock bursts, combined with field tests, selecting reasonable blasting parameters, including the charge diameters, the distances between holes, etc., the authors of this book apply the numerical simulation method to compare the data before and after blasting, with an aim to analyze the effectiveness of pre-relief de-stress controlled blasting and to verify the rationality of those blasting parameters. The main research contents in this book are as follows:

1) Study on the mechanism of pre-relief de-stress controlled blasting for high-intensity rockbursts

The research of the mechanism of de-stress controlled blasting for high-intensity rockbursts is the premise to determine the drilling and blasting parameters. Through deducing the theoretical formula of stress relief, the calculation formula of residual stress is obtained. The mechanism of rock bursts is mainly considered from two aspects: geostress and energy. Firstly, based on rock mass energy theory, we'll carry out some relevant field and laboratory experiments, the numerical simulation and inversion analysis, to reveal the interaction mechanism between pre-release de-stress blasting and geostress; thus we can lay a theoretical foundation for the pre-relief de-stress controlled blasting of high-intensity rock bursts. In order to avoid large-scale rock bursts after excavation, an artificial blasting is used to intervene to pre-release the stress. Therefore, using the controlled blasting method to release the stress of surrounding rocks before the tunnel excavation and to reduce the level of energy accumulated can reduce or eliminate the hazard of high-intensity rock bursts.

2) Study on the optimization of blasting parameters for pre-relief de-stress controlled blasting

Based on the theory of de-stress controlled blasting, we eximine the effects of various rock parameters, geometric sizes and excavation conditions on stress relief. Through the simulation of ANSYS and FLAC software and the orthogonal analysis, the value range of single parameter is optimized. After comprehensive simulations of the additional deformation and the stress field disturbance caused by many factors, an optimized combination of parameters is obtained.

3) Field tests and verification of pre-relief de-stress controlled blasting

Due to the high-frequency of strong rock bursts, the high-stress tunnel section in Sangzhuling tunnel of Linzhi-Lasha Railway is selected as test site to carry out the pre-relief de-stress controlled blasting. Based on the optimized blasting parameters, we apply the stress relief method to test the stress data before and after blasting. By comparing those data before and after blasting, we try to verify the rationality of the numerical simulation results. Then, the residual stress formula of de-stress blasting is derived and used in calculation, the result of which is compared with the field test results so as to verify each other.

1.3.2 Technical route

Based on a large number of relevant literature, comprehensivly applying the theories of rock mechanics and blasting dynamics, the authors conduct a research on the mechanism of pre-release de-stress controlled blasting for high-intensity rock bursts. After surveying many rockburst tunnels under construction, the researching group is determined to choose the high-rockburst-intensity section in Yakang High-speed Erlangshan tunnel, the section in Sangzhuling tunnel of Linzhi-Lhasa Railway and other tunnels with high geostress as the experimental sites, and collects large amount of rock samples from deep surrounding rocks, and take them back for laboratory tests, with an aim to obtain the most effective mechanical parameters, such as the compressive strength of rocks.

After testing the geostress in the sampling sites, the initial stress value and the stress value after excavation are obtained. Based on theoritical analysis, two kinds of parameters: the rock mechanical parameters (main mechanical parameters) and the construction conditions (excavation location and excavation size) are analyzed to build up the rock stress models, which are input into the numerical simulation platform for interactive numerical simulation calculation. Then based on large-scale adjustment of model parameters, and a variety of different simulation conditions, a rock burst simulation platform is built, which can reflect the changing rules and scales of rock bursts under the conditions of different rock lithology parameters, different constructing conditions in high-intensity geostress area.

After investigating and summarizing the existing rock burst records and data, we can obtain some control parameters of high-intensity rock bursts, which, combined with the common geological conditions and the construction conditions, are put into the simulation platform. According to the actual situation, the rock burst processes in the simulation platform are continuously adjusted and the optimal simulated controlled blasting parameters can be obtained more accurately, through continuous comparison of the simulation results. Thus different geological conditions and corresponding controlled parameters are stored to form an adaptive database, and then the numerical model of pre-relief de-stress controlled blasting is established.

Firstly, a numerical model is established to simulate the effects of various factors on rock bursts, including lithology and excavation conditions, etc. Using the SPSS software to analyze the main influencing factors and get the influencing rules, and then, based on the results, some effective control ways for high-intensity rock bursts can be found. Then through formula derivation, the distribution rules of residual stress after de-stress blasting are summarized, which may lay a theoretical foundation for the subsequent numerical simulation. After that, the ANSYS/LS-DYNA software is used to simulate the stress changes of surrounding rocks under the conditions of setting excavating and drill-burst parameters, and the data are imported into the FLAC 3D software to calculate the residual stress of surrounding rocks in the tunnel and obtain the distribution rules. In the numerical model of de-stress controlled blasting, the optimal blasting parameters are matched. Finally, the above results are applied to the prevention of rock bursts in those high-intensity sections, such as in Erlangshan Highway Tunnel of Sichuan or in Sangzhuling tunnel of Linzhi-Lasha Railway, so as to modify the value criteria of de-stress parameters, and to verify the correctness of the derived formula.

The specific technical route is shown in Figure 1.2.

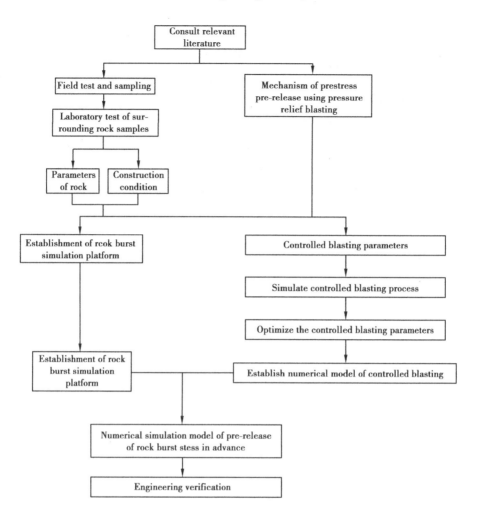

Figure 1.2　Schematic diagram of technical route

1.4　Expected Target

①Through the numerical simulation analysis, we try to find the main factors of rock burst intensity, discover its rules, and explore the best de-stress way for the high-intensity rock bursts.

②Based on theories of rock mechanics and of blasting mechanics, we try to deduce the calculation formula for residual geostress in de-stress blasting, and through analyzing the interaction mechanism between pre-release de-stress blasting and geostress, we can solve the technological problems in the follow-up study of the pre-relief de-stress controlled blasting for high-intensity rock blasting, and lay some theoretical foundation for the de-stress blasting in high-intensity rocks.

③Through numerical simulation calculation and field test analysis, we'll propose a set of optimized design criteria for the combination of parameters in the pre-release de-stress controlled blasting. By analyzing the mechanism, the process principle and the applicable scope of the pre-relief de-stress controlled blasting, we hope to form a complete set of techniques and methods for the pre-release de-stress controlled blasting in high-intensity rock bursts.

1.5 Main Innovations

Based on literature reviews and previous experience, the authors plan to apply these research methods: theoretical analysis, numerical simulation, field test verification, etc. The main innovations are as follows:

①Through theoretical analysis, we deduce the calculation formula of the residual geostress in the surrounding rocks after the pre-release de-stress blasting, and the field tests also verify the rationality of the formula, which provides a theoretical support for the evaluation of de-stress blasting effect.

②The numerical simulation software ANSYS/ LS-DYNA and FLAC 3D are used to simulate the de-stress blasting in surrounding rocks under high geostress. Multi-parameter orthogonal test and single factor variation are used to simulate the blasting scheme. According to the comprehensive evaluation of numerical simulation and the on-site blasting situation, the optimum technological parameters of de-stress blasting are determined, i.e., the length of the blasthole is 5 m, the spacing between blastholes is 70 cm, the uncoupling coefficient is 1.5, the diameter of blasthole is 50 mm, and the blasting mode is inverse initiation.

③Based on the results of theoretical analysis and numerical simulation, the pre-relief de-stress controlled blasting tests for high geostress were carried out in Sangzhuling tunnel of Linzhi-Lhasa Railway. The experimental results show that the maximum principal stress before blasting is 30.7 MPa, and the maximum principal stress after blasting is 11.5 MPa, which means the maximum principal stress is reduced by 63%. Therefore, the pre-relief de-stress controlled blasting has a significant effect on the reduction of the maximum principal stress. Thus the field tests verify that the pre-release de-stress controlled blasting method proposed in this study is an effective solution to the prevention and control of high-intensity rock bursts.

Chapter 2　Factors and Preventing Methods of Rockbursts

Rockburst is a geological threat induced by human engineering activities in high-stress surrounding rocks with sudden release of stress. It usually appears in the form of spalling, ejecting, or even a partial collapse in tunnels. The scale can be minor or severe. It is often accompanied by deformations of rocks or loud voices. The occurrence of rockbursts is affected by the mechanical properties of surrounding rocks and the excavation process, etc. Thus this chapter studies the effects of mechanical properties of rocks, the process of excavation and other factors on rockbursts, with an aim to analyze its effect rules and take reasonable de-stress measures.

2.1　Definition of Rockburst

A rockburst is the sudden and violent release of the energy stored in the rock mass under certain conditions, in the simultaneous forms of bursting or ejecting of rocks. Generally speaking, in a rock mass with high brittleness and elasticity, when the geostress exceeds the strength of the rock itself, or when some underground engineering activities disrupt the natural balance of rock mass, the released energy accumulated in the rock mass can lead to rock bursts, where rocks are destroyed and thrown out. In minor rock bursts, only rock fragments are peeled off without ejection phenomenon, but severe rock bursts can trigger an earthquake with 4.6 in magnitude and 7-8 degree in intensity, which may cause damage to the ground structure and make a loud noise. The rockbursts may occur suddenly and end quickly, but they may also last for a long time, ranging from days to months.

In the fileds of hydropower, transportation, tunnel engineering and other industries, it is generally called a rockburst, while in the coal mine, metallurgy and other industries, it is also called Rock Burst. There also exists difference in the forming mechanism between them. A rock burst is a type of rock mass failure, where a stable rock mass in the roadway or the tunnel face, under high geostress conditions, loses its original balance of geostress due to some engineering excavation, blasting, stress waves, etc., and begins to explode or catalyze, accompanied by sounds, in order to release the high energy accumulated in it and form a new state of stress balance. The percussive ground pressure is a disruptive mine dynamic phenomenon under high stress, where the ore rock mass around the mine lane or mining face are damaged due to the instant release of elastic strain energy, accompanied by a huge noise, with the ore rock mass thrown into mining space and some gas waves generated.

Currently, consensus has not yet been reached on the classification of rockbursts in the academic community. They are classified mainly by the location of the rockburst and the amount of released energy, the inherent stress factors, and the way the rockburst rock is destroyed.

A rockburst is the result of combined action of multiple factors. Each factor affects a rockburst sometimes jointly with other factors, sometimes in an independent manner, and each factor has different degree of effect on the occurrence of a rockburst. As a necessary factor for the occurrence of rockbursts, geostress in rock mass can be generated and accumulated in the following ways: plate boundary compression, mantle thermal convection, geocentric gravity, internal earth stress, etc. In addition, the unevenness in temperature, the gradients of water pressure, the erosion on surface land or other physical and chemical changes can also cause changes in geostress. The tectonic stress field and the gravity stress field are the main components of the rockburst geostress field. Some recent tectonic activities have caused such high geostress in mountains, that the hard and complete rocks with high brittleness and elasticity, even with few or only one crack, will accumulate a large amount of strain energy.

2.2 Conventional Evaluation Indexes of Rockbursts

In order to predict a rockburst and its possible intensity to take corresponding preventive measures, a lot of researches have been done by scholars at home and abroad. Currently, the main criteria for rockburst are as follows.

1) E. HOKE method

This method is to use the ratio of the maximum tangential stress in tunnel face to the uniaxial compressive strength of rocks to determine the intensity of the rockburst, and to take corresponding measures.

$$\sigma_{max}/\sigma_c = \begin{cases} 0.34 & \text{(A few tablets, Grade I)} \\ 0.42 & \text{(Severe Gangbang, Grade II)} \\ 0.56 & \text{(Heavy support, class III)} \\ > 0.70 & \text{(Severe rockburst, class IV)} \end{cases} \quad (2.1)$$

Where σ_{max} is maximum tangential stress in tunnel face, and σ_c is rock uniaxial compressive strength.

2) Turchaninov method

Based on his mine construction experience, Turchaninov proposed the ratio of the sum of the tangential stress $\sigma_{\theta max}$ and the axial stress σ_L of the rock burst to the uniaxial compressive strength σ_c of rocks as follows:

$$\begin{cases} (\sigma_{\theta max} + \sigma_L)/\sigma_c \leq 0.3 & (\text{No rock burst}) \\ 0.3 < (\sigma_{\theta max} + \sigma_L)/\sigma_c \leq 0.5 & (\text{Possible rock burst}) \\ 0.5 < (\sigma_{\theta max} + \sigma_L)/\sigma_c \leq 0.8 & (\text{Must. be burst}) \\ 0.8 < (\sigma_{\theta max} + \sigma_L)/\sigma_c & (\text{Severe rockburst}) \end{cases} \quad (2.2)$$

3) Kidybinski method

Scholars have proposed the W_{et} criterion for the elastic energy (strain energy) index based on coal tests, W_{et} is the ratio of elastic strain energy to loss strain energy, that is,

$$W_{et} = \varphi_{sp}\varphi_{st} \quad (2.3)$$

Among them: φ_{sp}, φ_{st} are the elastic strain energy and loss strain energy of the test block, both being calculated from the area of the test block for loading and unloading stress-strain curve. The criteria of W_{et} are as follows:

$$\begin{cases} W_{et} \geq 5 & (\text{Severe rockburst}) \\ W_{et} = 2.0 - 4.9 & (\text{Medium rockburst}) \\ W_{et} < 2.0 & (\text{No rock burst}) \end{cases} \quad (2.4)$$

4) Russense criterion

The Russense criterion method is to establish a relationship diagram for rockburst intensity, based on the relationship between the maximum tangential stress σ_θ of the cave and the point loading strength of rock I_s. When the point loading strength I_s is converted into the uniaxial compressive strength σ_c of rock, we can use the rockburst intensity relationship diagram to judge the possibility of rockburst. The judgement is as follows:

$$\begin{cases} \sigma_\theta/\sigma_c < 0.20 & (\text{No rock burst}) \\ 0.20 \leq \sigma_\theta/\sigma_c < 0.30 & (\text{Weak rockburst}) \\ 0.30 \leq \sigma_\theta/\sigma_c < 0.55 & (\text{Medium rockburst}) \\ 0.55 \leq \sigma_\theta/\sigma_c & (\text{Severe rockburst}) \end{cases} \quad (2.5)$$

5) Erlangshan tunnel criteria

Xu Linsheng and Wang Lansheng improved the criteria based on more than 200 rockburst data items recorded during the construction of Erlangshan Highway Tunnel. At the site, a point loading meter was used to determine the point loading strength of rock $I_{s(50)}$. Then through the formula: $\sigma_c = 22I_{s(50)}$, the uniaxial compressive strength of rock σ_c can be calculated. After that, an improved on-site stress recovery test method was used to test the readings of the point loading meter at the strain recovery time of rock core in the shallow surface of the cave wall F (MPa). Then the following formula can be used to find σ_θ:

$$\sigma_\theta = (FS_p a)/(LH) \tag{2.6}$$

Among them: S_p is the piston area of point loading meter jack (15.5 cm^2), a is the equivalent coefficient (1.324), L is the chord length (3.3 cm), and H is the core sample length (cm).

Based on the test results and on-site measures of surrounding rocks, Linsheng Xu and Lansheng Wang made the following rockburst classifications:

$$\begin{cases} \sigma_\theta/\sigma_c < 0.3 & (\text{No rock burst}) \\ 0.3 \leq \sigma_\theta/\sigma_c < 0.5 & (\text{Minor rockburst}) \\ 0.5 \leq \sigma_\theta/\sigma_c < 0.7 & (\text{Medium rockburst}) \\ 0.7 \leq \sigma_\theta/\sigma_c & (\text{Severe rockburst}) \end{cases} \tag{2.7}$$

6) Critical buried depth criterion

Although the critical buried depth is not the only trigger for the occurrence of rockbursts, when not considering the effect of tectonic stress, it is very applicable to the rockmass with high self-gravity stress. As long as the thickness of the overlying stratum is greater than the critical buried depth calculated by the formula, a rock burst will occur. The criteria are as follows:

$$H_{cr} = \frac{0.318(1-\mu)}{(3-4\mu)\gamma}\sigma_c \tag{2.8}$$

Where μ is the poisson's ratio of rock, γ is the bulk density of rock mass, and σ_c is the uniaxial compressive strength of rock mass.

7) Zhenyu Tao's Criterion and his rockburst classification

Based on the previous research and domestic engineering experience, Zhenyu Tao put forward that no rock burst will happen if $\sigma_c/\sigma_1 > 14.5$, and a rockburst might happen if $\sigma_\theta/\sigma_1 \leq 14.5$. He also divided rockbursts into 4 levels, as shown in Table 2.1. (σ_c is the uniaxial compressive strength of rock; and σ_1 is the maximum principal stress).

Table 2.1 Classification of rockbursts proposed by Zhenyu Tao

Rockburst classification	σ_c/σ_1	Description
I	>14.5	No rock burst and no acoustic emission
II	14.5~5.5	Low rockburst activity with slight acoustic emission
III	5.5~2.5	Medium rockburst activity with strong acoustic emission
IV	<2.5	High rock burst activity with loud cracking sound

The above criteria have their distinct characteristics and can be used in practice according to the actual conditions of the project.

2.3 Establishment of Numerical Model for Analyzing Factors of Rockbursts

Characterized by suddenness and high risk, rockbursts are extremely difficult to be observed on site and tested in physical simulation. Under this limitation, we can use some software to numerically simulate the effects of the physical properties of rocks, the construction techniques, and the supporting structures, etc. so as to find ways to control rockbursts. The FLAC 3D software can be used to simulate the mechanical properties of three-dimensional rocks and other materials, especially of the plastic rheological properties when the yield limit is reached; thus it is widely used in the construction of slopes, underground caverns and other fields. Therefore, FLAC 3D software is selected as the main tool to study the main factors of rockbursts.

2.3.1 Introduction to simulation tools

FLAC, the Shorthand of Fast Lagrangian Analysis of Continua, was first used by Willkins in the field of solid mechanics. With the strong promotion of American ITAXCA Consulting Group in the world, the FLAC 3D software has become an important numerical simulation tool in geotechnical calculations. Including constitutive models of 11 materials, FLAC 3D has a wide range of applications. There are 5 calculation modes in this software: the static mode, dynamic mode, creep mode, percolation mode, and temperature mode, among which the calculations of temperature can be coupled to the static, the dynamic or the percolation, and they can also be calculated separately.

FLAC 3D can simulate a variety of structures:

①For ordinary rocks, soil, or other matters, it provides an eight-node hexahedral unit simulation.

②The FLAC 3D grid has interfaces, which can divide the computing grid into several parts. The grids on two sides of the interface can be separated or slide. So the interface is used to simulate the joints and faults or virtual physical boundaries.

③FLAC 3D contains four structural units: the beam unit, the anchor unit, the pile unit, and the shell unit, all of which can be used to simulate the artificial structures in geotechnical projects, such as the supporting structure, the lining, the anchor cable, the rock bolt, the geotextile, and the friction piles and sheet piles. FLAC 3D has multiple boundary conditions.

Of course, FLAC 3D also has certain limitations. For linear issues, FLAC 3D takes more calculation time than the corresponding finite element. Therefore, FLAC 3D is more effective in simulating nonlinear issues, large deformation issues, or dynamic issues. Furthermore, the convergence speed of FLAC 3D depends on the maximum natural period of the system.

2.3.2 Model establishment and benchmark simulation

Those high-geostress and rockburst-prone areas are mostly dominated by those horizontal tectonic stresses, widely distributed in southwest China. Rockburst-prone areas generally exist in those deep-buried limestone, granite, quartz sandstone, andesite, and diorite. This model takes a rockburst tunnel in southwest China as an example to explore the effects of the physical and mechanical properties of surrounding rocks on the rockbursts. The surrounding rocks in the tunnel are andesite. According to the literature, the physical and mechanical parameters of andesite are in wide scopes: the distribution range of compressive strength: 80-250 MPa, the distribution range of tensile strength: 10-20 MPa, the distribution range of elastic modulus: 8-120 GPa, the distribution range of internal friction angle: 25°-60°, the cohesion distribution range: 5-40 MPa, the coisson's ratio distribution range: 0.18-0.35, and the density distribution range: 2,500-2,700 kg/m³.

In the simulated tunnel, the length is 50 m, the buried depth is 1,000 m, the hole diameter is 10 m, the horizontal geostress is 50 MPa, and the vertical geostress is 32 MPa. 100 m below the center axis of the tunnel is the fixed boundary, and the 50 m to the left and right of the center axis is the free boundary. The data in the following table is used as the reference value for simulation.

The model and mesh of simulated tunnel are shown in the following Figure 2.1.

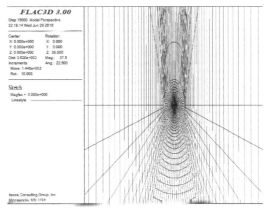

Figure 2.1　Tunnel model

After setting the corresponding parameters according to Table 2.2, the initial geostress is generated.

Table 2.2　Reference table of physical and mechanical parameters of andesite

σ_c/MPa	E/GPa	ν	$\varphi/(°)$	C/MPa
85	28	0.19	28	5

After careful examination of the above figures, it is found that the initial geostress value simulated by the software is similar to the given initial geostress. In order to ensure the reliability of the model, the maximum imbalance curve is taken. From the Figure 2.2 and Figure 2.3, we find that when the calculation reaches 2,000 steps, the maximum imbalance force starts to converge, indicating that the model results are reliable.

Then we begin to simulate full face excavation. When we excave for 5 m, the horizontal and vertical geostress cloud diagrams are shown as Figure 2.4–Figure 2.6.

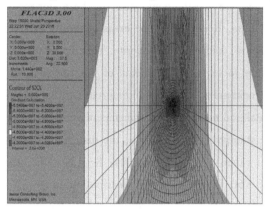
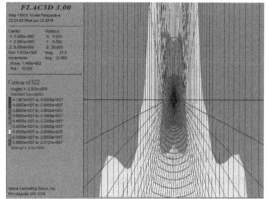

Figure 2.2　Horizontal initial geostress　　　　　　Figure 2.3　Vertical initial geostress

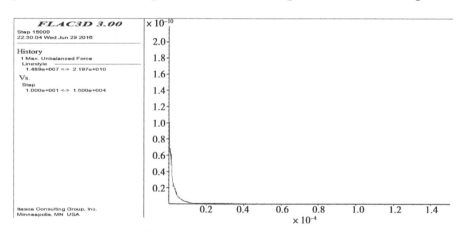

Figure 2.4　Maximum unbalanced force curve

As can be seen from the above figures, after excavation, the surrounding rocks in the middle and upper part of the tunnel form a stress concentration area, which is the rockburst-prone area. According to the tangential principal stress formula (2.9):

$$\sigma_\theta = \sigma_1 + \sigma_3 + 2 \times (\sigma_1 - \sigma_3 \cos 2\theta) \tag{2.9}$$

The rockburst judging coefficient is obtained: $\sigma_\theta / \sigma_c = 1.3031$.

According to the tunnel rockburst criteria proposed by Lansheng Wang and Linsheng Xu, we find this section of tunnel is an area higly prone to rockbursts. Then based on this section, we carry out the numerical simulations of the effects of such factors on rockbursts as the physical and mechanical properties of surrounding rocks, the tunnel construction techniques, and the tunnel supporting structures, ect.

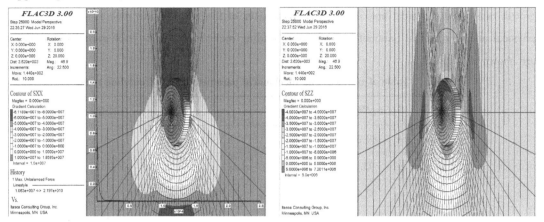

Figure 2.5　Horizontal geostress cloud diagram after excavation

Figure 2.6　Vertical geostress cloud diagram after excavation

2.4　Impact of Rock's Physical and Mechanical Properties on Rockbursts

2.4.1　Qualitative analysis of the impact of rock's physical and mechanical properties on rockbursts

The statistical function is the core part of SPSS (Statistical Product and Service Solutions). The basic statistical functions include: the description and preprocessing of sample data, the hypothesis testing, the analysis of variance, the regression analysis, the log-linear analysis, the judge analysis, the factor analysis, the correspondence analysis, the birth and death analysis, and the reliability analysis. There are many influencing factors and many variables in analysis, all of which will participate in data modeling. Thus there are large calculation workloads in the analysis process. Meanwhile, the high overlaps and high correspondents of information between variables pose obstacles to the application of statistical methods. Therefore, the factor analysis in this powerful statistical software SPSS is necessary for data processing. In order to explore the impact of rock's physical and mechanical properties on rockbursts, multiple sets of simulations are performed on the basis of the reference value. The specific simulation scheme is as follows: increase one of the parameters in the reference value and keep the remaining parameters unchanged, then make simulations and, by using the principal stress formula, calculate the corresponding σ_θ/σ_c after the increased parameter. We need first to determine the mechanical parameter intervals of the surrounding rocks when a rock burst occurs, and the growth rate of each physical mechanical parameter. For the convenience of simulation calculation, the growth rate of the physical and mechanical parameter of the surrounding rock is set to 5%. The physical and mechanical parameters of the surrounding rocks required for each group of simulation are shown in Table 2.3.

Table 2.3 Table of physical and mechanical parameters of surrounding rocks

	σ_c/MPa	E/GPa	ν	φ/(°)	C/MPa
Group 1	89.25	18	0.21	32	10
Group 2	93.5	28	0.23	36	15
Group 3	97.45	38	0.25	40	20
Group 4	102	48	0.27	44	25
Group 5	106.25	58	0.29	48	30
Group 6	110.5	68	0.31	52	35

In order to ensure that the sample increases at a rate of 5% in the reference value, the parameters are now modified as:

$$E_M = E + 192, \nu_M = \nu + 0.21, \Phi_M = \varphi + 52, C_M = C + 95$$

Table 2.4 shows the modified values of the corresponding parameters after the compressive strength of the sample increases at a rate of 5%. According to the physical and mechanical parameters of the surrounding rocks in Table 2.3 and Table 2.4, the tangential principal stress formula is used to calculate the σ_θ/σ_c as:

Table 2.4 Table of modified physical and mechanical parameters of rocks

	σ_c/MPa	E_f/GPa	ν_f	φ_f/(°)	C_f/MPa
Base value	85	200	0.4	80	100
Group 1(0.05)	89.25	210	0.42	84	105
Group 2(0.10)	93.5	220	0.44	88	110
Group 3(0.15)	97.45	230	0.46	92	115
Group 4(0.20)	102	240	0.48	96	120
Group 5(0.25)	106.25	250	0.52	100	125
Group 6(0.30)	110.5	260	0.54	104	130

By grouping the stress intensity ratios of the varying parameters in Table 2.5, according to the growth rate of each parameter, we can obtain Table 2.6.

Table 2.5 Table of physical and mechanical parameters of surrounding rocks and their corresponding stress-intensity ratios

σ_θ/MPa	E/GPa	ν	φ/(°)	C/MPa	σ_θ/σ_c
89.25	28	0.19	28	5	1.2411
93.5	28	0.19	28	5	1.1847
97.45	28	0.19	28	5	1.1332
102	28	0.19	28	5	1.086
106.25	28	0.19	28	5	1.0425
110.5	28	0.19	28	5	1.0024
85	38	0.19	28	5	1.3038
85	48	0.19	28	5	1.3122
85	58	0.19	28	5	1.3038
85	68	0.19	28	5	1.3042
85	28	0.21	28	5	1.2906
85	28	0.23	28	5	1.2773
85	28	0.25	28	5	1.272
85	28	0.27	28	5	1.2642
85	28	0.29	28	5	1.2558
85	28	0.31	28	5	1.2505
85	28	0.19	32	5	1.3015
85	28	0.19	36	5	1.2481
85	28	0.19	40	5	1.2193
85	28	0.19	44	5	1.2092
85	28	0.19	48	5	1.2014
85	28	0.19	52	5	1.1976
85	28	0.19	28	10	1.1931

Continued

σ_θ/MPa	E/GPa	ν	$\varphi/(°)$	C/MPa	σ_θ/σ_c
85	28	0.19	28	15	1.1936
85	28	0.19	28	20	1.1899
85	28	0.19	28	25	1.1891
85	28	0.19	28	30	1.1884
85	28	0.19	28	35	1.1884

Table 2.6　Stress intensity ratio table

	σ_c/MPa	E_f/GPa	ν_f	$\varphi_f/(°)$	C_f/MPa
Group 1(0.05)	1.241100	1.305700	1.290600	1.301500	1.193100
Group 2(0.10)	1.184700	1.306700	1.277300	1.248100	1.193600
Group 3(0.15)	1.133200	1.306200	1.272000	1.219300	1.189900
Group 4(0.20)	1.086000	1.313800	1.264200	1.209200	1.189100
Group 5(0.25)	1.042500	1.307200	1.255800	1.201400	1.188400
Group 6(0.30)	1.002400	1.306300	1.250500	1.197600	1.188400

Using the SPSS software to make factor analysis of data in Table 2.6, we obtain the correlation matrix (Table 2.7), and the table of factor eigenvalues and contribution rate (Table 2.8).

Table 2.7　Correlation matrix

	σ_c/MPa	E_f/GPa	ν_f	$\varphi_f/(°)$	C_f/MPa
σ_c	1.000	−0.195	0.995	0.927	0.917
E_f	−0.195	1.000	−0.173	−0.276	−0.302
ν_f	0.995	−0.173	1.000	0.939	0.884
φ_f	0.927	−0.276	0.939	1.000	0.867
C_f	0.917	−0.302	0.884	0.867	1.000

Table 2.8 Factor eigenvalues and contribution rates

Factor	Initial eigenvalue		
	Eigenvalues	Contribution rate/%	Cumulative contribution rate/%
1	3.844	76.889	76.889
2	0.938	18.754	95.643
3	0.142	2.840	98.484
4	0.075	1.500	99.984
5	0.001	.016	100.000

Use the SPSS software to calculate the eigenvalues of the correlation matrix, which are shown in Table 2.8. The first three eigenvalues in the matrix are $\lambda_1 = 3.884$, $\lambda_2 = 0.938$, and $\lambda_3 = 0.142$. The cumulative contribution rate of the eigenvalues to the sample variance is $(\lambda_1 + \lambda_2 + \lambda_3)/5 = 0.984$, which meets the principle that the cumulative contribution rate of the extracted common factor is no less than 85%; so the number of common factors is selected to be 3.

It can be seen from Figure 2.7 that the eigenvalue corresponding to factor 3 is an obvious inflection point. This point is preceded by a steep polyline connected to a large factor and then followed by a gentle polyline connected to a small factor. As can be seen from Figure 2.7 Crushed stone, the eigenvalues correspond to the first two factors are larger and then they fall rapidly, while the latter three eigenvalues are smaller, and the corresponding polyline is smoother, which indicates that factor 1 and factor 2 are the main influencing factors.

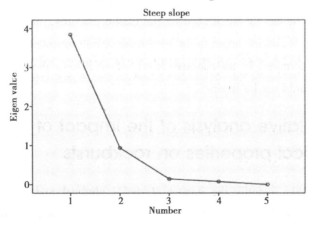

Figure 2.7 Crushed stone

Now let's rotate the factor load matrix and get Table 2.9, which is the factor loading matrix after the orthogonal rotation with maximum variance.

Table 2.9 Factor loading matrix after rotation

	Factor				
	1	2	3	4	5
σ_c	0.978	−0.106	0.171	−0.056	0.022
E	−0.133	0.989	−0.057	−0.019	0.000
ν	0.990	−0.101	0.090	−0.023	−0.019
φ	0.923	−0.219	0.104	0.297	−0.001
C	0.830	−0.196	0.521	0.031	0.000

Based on the rotated factor loading matrix, the factor expressions for each observed variable can be written as follows:

$$\sigma_{cf} = 0.978F_1 - 0.106F_2 + 0.171F_3$$
$$E_f = -0.133F_1 + 0.989F_2 - 0.057F_3$$
$$\nu_f = 0.99F_1 - 0.101F_2 + 0.09F_3$$
$$\varphi_f = 0.923F_1 - 0.219F_2 + 0.104F_3$$
$$C_f = 0.83F_1 - 0.196F_2 + 0.521F_3$$

According to the factor analysis model, the first main factor F_1 is mainly determined by two indexes: the compressive strength and the internal friction angle, whose loan on the main factor F_1 is above 0.9, which shows the responsiveness of rock's intensity to changes of rockbursts. Moreover, the main factor F_1's contribution of variance of the variable has reached 76.889%, indicating that F_1 is an important aspect in the responding system of rockburst activities; the second main factor F_2 is mainly determined by the elastic modulus, which shows that the change of the elastic modulus is also important to the responsiveness to rockburst activities; the third main factor F_3 is mainly determined by the cohesive force, whose loading index on the main factor F_3 is only 0.521, but it should not be ignored.

2.4.2 Quantitative analysis of the impact of rock's physical and mechanical properties on rockbursts

In order to further quantify the impact of rock's physical and mechanical properties on rockbursts and to ensure that the parameters of rock's physical and mechanical properties fall within a reasonable interval, we apply the least square method to make calculations and finally determine the values of F_1, F_2 and F_3 are 0.2209, 0.5924 and 0.3376 respectively. Then we amplify some of the coefficients in the factor expressions of each observed variable in 2.4.1.

$$\sigma_c = 3.912F_1 - 0.424F_2 + 0.684F_3$$
$$E = -0.133F_1 + 0.989F_2 - 0.057F_3$$
$$N = 0.99F_1 - 0.101F_2 + 0.09F_3$$
$$\varphi = 2.769F_1 - 0.657F_2 + 0.612F_3$$
$$C = 0.83F_1 - 0.196F_2 + 0.521F_3$$

Because the three major factors are orthogonal to each other, the correlation between the three major factors is 0, which means that the three major factors are independent from and do not interfere with each other. When F_2 and F_3 remain unchanged and only F_1 is changed, we discuss the responsiveness of rockburst to F_1, that is, the changes of the stress intensity ratio. According to the above formula, when F_2 and F_3 remain unchanged and only F_1 is changed, the corresponding physical and mechanical parameter values are calculated. Then these values are input into the numerical model and, by using the tangential principal stress formula, the corresponding stress intensity ratios are calculated. The specific values are shown in the following Table 2.10.

Table 2.10 F_1 and stress intensity ratio

F_1	σ_c	E	ν	ψ	C	K	G	σ_θ	Strength-stress ratio
0.2209	84.39	53.73	0.1892	42.91	24.31	28.8127	22.5908	64.6	0.7654
0.2309	88.30	53.59	0.1991	45.68	25.14	29.6831	22.3459	64.5	0.7304
0.2409	92.21	53.46	0.209	48.45	25.97	30.6185	22.1091	64.5	0.6994
0.2509	96.13	53.33	0.2189	51.21	26.8	31.6198	21.8762	63.3	0.6584
0.2609	100.04	53.19	0.2288	53.98	27.63	32.6880	21.6430	63.3	0.6327
0.2709	103.95	53.06	0.2387	56.75	28.46	33.8436	21.4176	63.7	0.6127
0.2809	107.86	52.93	0.2486	59.52	29.29	35.0901	21.1957	63.6	0.5896
0.2909	111.77	52.8	0.2585	62.29	30.12	36.4389	20.9773	63.5	0.5681

Using SPSS software to perform curve fitting on F_1 and stress intensity ratio, we'll get each curve fitting model, shown in the following Figure 2.8.

As can be seen from the figure above, no matter which curve is used to fit the changing law between F_1 and the stress intensity ratio, the general trend is consistent. As F_1 increases, the stress intensity ratio gradually decreases, which means that with the increase of F_1, rockburst activities gradually weaken. Through careful observation and comparison of the curve model in the figure above, we find that the quadratic curve model is the most consistent, so we choose the quadratic curve to fit the changing law between F_1 and the stress intensity ratio.

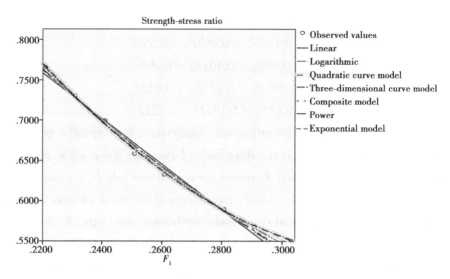

Figure 2.8 Curve fitting model

$$\sigma_\theta/\sigma_c = 2.3945 - 10.8195F_1 + 15.6245F_1^2 \quad (2.10)$$

When F_1 and F_3 remain unchanged and only F_2 is changed, we discuss the responsiveness of rockburst to F_2, that is, the change of the stress intensity ratio. According to the above formula, when F_1 and F_3 remain unchanged and only F_2 is changed, the corresponding values of each physical and mechanical parameter are calculated. These values are then input into a numerical model and, by using the tangential principal stress formula, the corresponding stress intensity ratios are calculated. The specific values are shown in the following Table 2.11.

Table 2.11 F_2 and stress intensity ratio table

F_2	σ_c	E	ν	ψ	C	K	G	σ_θ	Strength-stress ratio
0.5924	84.39	53.73	0.1892	42.91	24.31	28.81274	22.59082	64.6	0.76549354
0.6224	83.12	56.69	0.1862	40.94	23.72	30.10941	23.89563	64.1	0.77117421
0.6524	81.85	59.66	0.1832	38.97	23.14	31.38678	25.21129	64.2	0.78436164
0.6824	80.57	62.63	0.1802	36.99	22.55	32.64019	26.53364	64	0.79434033
0.7124	79.3	65.59	0.1771	35.02	21.96	33.85465	27.86084	64	0.80706179
0.7424	78.03	68.56	0.1741	33.05	21.37	35.06188	29.19683	64	0.82019736
0.7724	76.76	71.53	0.1711	31.08	20.78	36.24709	30.53966	64.2	0.83637311
0.8024	75.49	74.5	0.168	29.11	20.2	37.3996	31.89212	63.9	0.84646973

Using SPSS software to perform curve fitting on F_2 and stress intensity ratio, we'll get each curve fitting model, shown in the following Figure 2.9.

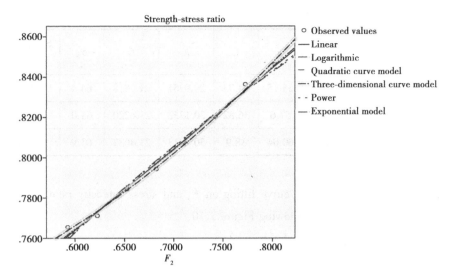

Figure 2.9 Curve fitting model

It can be seen from the above figure that no matter which curve is used to fit the changing law between F_2 and the stress intensity ratio, the general trend is consistent. As F_2 increases, the stress intensity ratio also gradually increases. This also means that with the increase of F_1, rockburst activities gradually increase. Through careful observation and comparison of the curve models in the figure above, we find that the cubic curve model is the most consistent, so the cubic curve is selected to fit the changing law between F_2 and stress intensity ratio.

$$\sigma_\theta / \sigma_c = 0.6881 + 0.1219 F_2^2 + 0.1582 - 1 \qquad (2.11)$$

When F_1 and F_2 remain unchanged and only F_3 is changed, the response of rockburst to F_3, that is, the change of the stress intensity ratio is discussed. According to the above formula, when F_1 and F_2 remain unchanged and only F_3 is changed, the corresponding rock's physical and mechanical parameter values are calculated. Then these values are input into the numerical model and, by the tangential principal stress formula, the corresponding stress intensity ratios are calculated. The specific values are shown in the following Table 2.12.

Table 2.12 F_3 and stress intensity ratio table

F_3	σ_c	E	v	ψ	C	K	G	σ_θ	Strength-stress ratio
0.3376	84.39	53.73	0.1892	42.91	24.31	28.8127	22.5908	64.6	0.7654
0.3776	87.13	53.5	0.1928	45.36	26.4	29.0256	22.4262	64.5	0.7402
0.4176	89.86	53.27	0.1964	47.8	28.48	29.2435	22.2626	64.8	0.7211
0.4576	92.60	53.04	0.2000	50.25	30.56	29.4666	22.1000	64.4	0.6954
0.4976	95.33	52.81	0.2036	52.7	32.65	29.6952	21.9383	64.2	0.6734

Continued

F_3	σ_c	E	v	ψ	C	K	G	σ_θ	Strength-stress ratio
0.5376	98.07	52.59	0.2072	55.15	34.73	29.9351	21.7818	64.4	0.6566
0.5776	100.81	52.36	0.2108	57.6	36.82	30.1752	21.6220	64.0	0.6348
0.6176	103.54	52.13	0.2144	60.04	38.9	30.4213	21.4632	63.9	0.6171

Using SPSS software to perform curve fitting on F_3 and stress intensity ratio, we'll get each curve fitting model, shown in the following Figure 2.10.

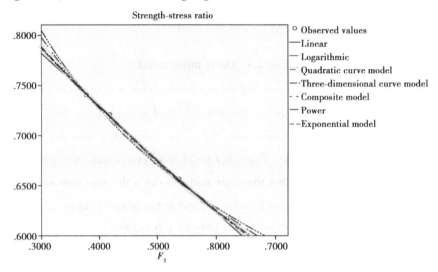

Figure 2.10 **Curve fitting model**

As can be seen from the Figure above, no matter which curve is used to fit the changing law between F_3 and the stress intensity ratio, the general trend is consistent. As F_3 increases, the stress intensity ratio gradually decreases. This also means that with the increase of F_3, rockburst activities gradually weaken. Through careful observation and comparison of the curve models in the figure above, we find that the quadratic curve model is the most appropriate, so the quadratic curve is selected to fit the changing law between F_3 and the stress intensity ratio.

$$\sigma_\theta/\sigma_c = 1.0031 - 0.799F_3 + 0.2816F_3^2 \qquad (2.12)$$

The above analysis shows that the intensity of rockbursts can be reduced, by changing the compressive intensity and the internal friction angle of the rock mass in the stress redistribution area to affect F_1, by changing the elastic modulus of the rock mass in the stress redistribution area to affect F_2, and by changing the viscosity of the rock mass in the stress redistribution area to affect F_3.

2.5 Numerical Analysis of the Effects of Excavating and Supporting Conditions on Rockbursts

According to rock mechanics, for the same geotechnical material, the geotechnical strength under different loading and unloading conditions is also different, so different excavating conditions will also affect the intensity of rockbursts. Thus, in this part, we try to establish a numerical model, as described in Section 2.3, to explore the impact of excavating conditions on rockbursts.

2.5.1 Numerical analysis of the effects of full-face excavation and stepwise excavation on rockbursts

In full-face exavation, the impact of tunnel sizes on rockbursts is simulated, and after consulting relevant literature, the changing range of tunnel radius. In order to depict accurately the impact of the excavation sizes on the intensity of rockburst activity, as many simulation groups as possible are needed; thus we finally determine that the tunnel radius is increased by a gradient of 0.2 m. Then we input the radius of each tunnel into the numerical model to get the corresponding horizontal and vertical stresses. After substituting them into the tangential principal stress formula, we calculate the corresponding stress-intensity ratios. The specific results are shown in Table 2.13.

Table 2.13 Tunnel size and stress intensity ratio table

σ_c	σ_t	E	ν	K	G	φ	C	R	Strength-stress ratio
100.000	10.000	38.500	0.250	25.667	15.400	40.000	20.000	5.000	1.160
100.000	10.000	38.500	0.250	25.667	15.400	40.000	20.000	5.200	1.150
100.000	10.000	38.500	0.250	25.667	15.400	40.000	20.000	5.400	1.130
100.000	10.000	38.500	0.250	25.667	15.400	40.000	20.000	5.600	1.120
100.000	10.000	38.500	0.250	25.667	15.400	40.000	20.000	5.800	1.100
100.000	10.000	38.500	0.250	25.667	15.400	40.000	20.000	6.000	1.080
100.000	10.000	38.500	0.250	25.667	15.400	40.000	20.000	6.200	1.060
100.000	8.500	38.500	0.250	25.667	15.400	40.000	20.000	6.400	1.050
100.000	9.000	38.500	0.250	25.667	15.400	40.000	20.000	6.600	1.030

Use the SPSS software to fit the curve of the changes between the tunnel sizes and the stress intensity ratios, shown in the following Figure 2.11.

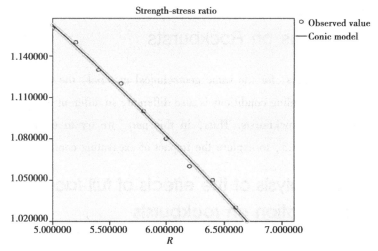

Figure 2.11 Quadratic curve fitting diagram

$$\frac{\sigma_\theta}{\sigma_c} = 1.4005 - 0.02056R - 0.00541R^2 \qquad (2.13)$$

In stepwise excavation, we try to establish a numerical model. We first excave the upper part of the tunnel. In order to eliminate the effects of some accidental factors on the numerical simulation results, multiple groups of simulations are performed here. Inputting the radius of each tunnel into the numerical model (by excavating only the upper half), we get the corresponding horizontal and vertical stresses. After substituting them into the tangential principal stress formula, we calculate the corresponding stress intensity ratios, which are shown in the following Table 2.14.

Table 2.14 Tunnel size and stress-intensity ratio table

σ_c	σ_t	E	ν	K	G	φ	C	R	Strength-stress ratio
100.000	10.000	38.500	0.250	25.667	15.400	40.000	20.000	5.000	1.030
100.000	10.000	38.500	0.250	25.667	15.400	40.000	20.000	5.200	1.010
100.000	10.000	38.500	0.250	25.667	15.400	40.000	20.000	5.400	0.990
100.000	10.000	38.500	0.250	25.667	15.400	40.000	20.000	5.600	0.970
100.000	10.000	38.500	0.250	25.667	15.400	40.000	20.000	5.800	0.964
100.000	10.000	38.500	0.250	25.667	15.400	40.000	20.000	6.000	0.953
100.000	10.000	38.500	0.250	25.667	15.400	40.000	20.000	6.200	0.942
100.000	8.500	38.500	0.250	25.667	15.400	40.000	20.000	6.400	0.928
100.000	9.000	38.500	0.250	25.667	15.400	40.000	20.000	6.600	0.913

SPSS software is used to fit the quadratic curve of the changes between the tunnel sizes and the stress intensity ratios, shown in the following Figure 2.12.

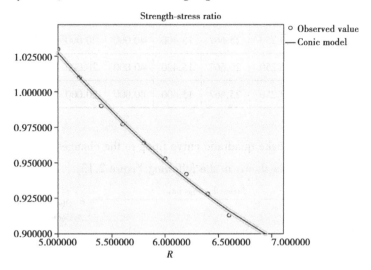

Figure 2.12 Quadratic curve fitting diagram

$$\frac{\sigma_\theta}{\sigma_c} = 1.753 - 0.20257R + 0.01147R^2 \tag{2.14}$$

Comparing Figure 2.11 with Figure 2.12, we find that as to the same tunnel size, the stress intensity ratio of the upper-part excavation is always much smaller than that of the full-face excavation.

Now after the excavation of the upper part of the tunnel and the redistribution of stress, the lower half of the tunnel is excavated. Inputting the radius of each tunnel into the numerical model, we obtain the corresponding horizontal and vertical stresses. After substituting them into the tangential principal stress formula, we calculate the corresponding stress-intensity ratios. The specific results are shown in the following Table 2.15.

Table 2.15 Tunnel size and stress intensity ratio table

σ_c	σ_t	E	v	K	G	φ	C	R	Strength-stress ratio
100.000	10.000	38.500	0.250	25.667	15.400	40.000	20.000	5.000	1.161
100.000	10.000	38.500	0.250	25.667	15.400	40.000	20.000	5.200	1.144
100.000	10.000	38.500	0.250	25.667	15.400	40.000	20.000	5.400	1.127
100.000	10.000	38.500	0.250	25.667	15.400	40.000	20.000	5.600	1.112
100.000	10.000	38.500	0.250	25.667	15.400	40.000	20.000	5.800	1.092
100.000	10.000	38.500	0.250	25.667	15.400	40.000	20.000	6.000	1.078

Continued

σ_c	σ_t	E	ν	K	G	φ	C	R	Strength-stress ratio
100.000	10.000	38.500	0.250	25.667	15.400	40.000	20.000	6.200	1.062
100.000	10.000	38.500	0.250	25.667	15.400	40.000	20.000	6.400	1.046
100.000	10.000	38.500	0.250	25.667	15.400	40.000	20.000	6.600	1.033

Use the SPSS software to make quadratic curve fiting to the changes between the tunnel sizes and the stress intensity ratios, as shown in the following Figure 2.13.

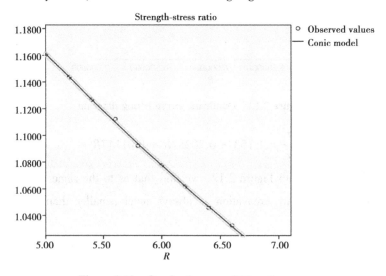

Figure 2.13 Quadratic curve fitting diagram

$$\frac{\sigma_\theta}{\sigma_c} = 1.7093 - 0.1314R + 0.004345R^2 \qquad (2.15)$$

Comparing Figure 2.11 with Figure 2.12 and Figure 2.13, we find that in stepwise excavation, the σ_θ/σ_c of the upper excavation is lower than that of full face excavation, and the/ σ_c of the lower excavation is a bit larger than that of the upper excavation, but still smaller than that of the full-face excavation. It shows that compared with full-face excavation, stepwise excavation helps to decrease rockburst activities.

2.5.2 Numerical analysis of the effect of supporting conditions on rockbursts

To explore the effect of supporting conditions on rockburst activities, we need to establish a numerical simulation model. Unlike 2.3.1, here we spray concrete to support the tunnel. In order to eliminate accidental factors affecting the numerical simulation results, multiple groups of simulations are needed. Inputting the radius of each tunnel into the numerical model, and performing a simple anchor spray support after the full face excavation, we can obtain the corresponding horizontal and vertical stresses. After substituting them into the tangential principal stress formula, we calculate the corresponding stress-intensity ratios. The specific results are shown in the following Table 2.16.

Table 2.16 Tunnel size and stress intensity ratio

σ_c	σ_t	E	ν	K	G	φ	C	R	Strength-stress ratio
100.000	10.000	38.500	0.250	25.667	15.400	40.000	20.000	5.000	1.159
100.000	10.000	38.500	0.250	25.667	15.400	40.000	20.000	5.200	1.138
100.000	10.000	38.500	0.250	25.667	15.400	40.000	20.000	5.400	1.125
100.000	10.000	38.500	0.250	25.667	15.400	40.000	20.000	5.600	1.107
100.000	10.000	38.500	0.250	25.667	15.400	40.000	20.000	5.800	1.091
100.000	10.000	38.500	0.250	25.667	15.400	40.000	20.000	6.000	1.075
100.000	10.000	38.500	0.250	25.667	15.400	40.000	20.000	6.200	1.063
100.000	10.000	38.500	0.250	25.667	15.400	40.000	20.000	6.400	1.048
100.000	10.000	38.500	0.250	25.667	15.400	40.000	20.000	6.600	1.032

Use the SPSS software to make quadratic curve fiting to the changes between the tunnel sizes and the stress intensity ratios, as shown in the following Figure 2.14.

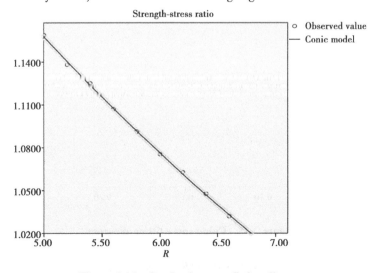

Figure 2.14 Quadratic curve fitting diagram

$$\frac{\sigma_\theta}{\sigma_c} = 1.7372 - 0.1447R + 0.005755R^2 \qquad (2.16)$$

After comparison, we find that after spraying concrete to support, σ_θ/σ_c is smaller than that of the unsupported σ_θ/σ_c.

In order to further explore the impact of the supporting thickness on the intensity of rockbursts, we make some corresponding adjustments based on the above numerical model. After inputting the figures of concrete-spray thickness into the numerical model, and performing some simple anchor spray support in the full-face excavation, we obtain the corresponding horizontal and vertical stresses. After substituting them into the tangential principal stress formula, we calculate the corresponding stress-intensity ratios. The specific results are shown in the following Table 2.17.

Table 2.17 Concrete spray thickness and stress intensity ratio

σ_c	σ_t	E	v	K	G	φ	C	h	Strength-stress ratio
100.000	10.000	38.500	0.250	25.667	15.400	40.000	20.000	0.500	1.1573
100.000	10.000	38.500	0.250	25.667	15.400	40.000	20.000	0.100	1.1579
100.000	10.000	38.500	0.250	25.667	15.400	40.000	20.000	0.150	1.1580
100.000	10.000	38.500	0.250	25.667	15.400	40.000	20.000	0.200	1.1588

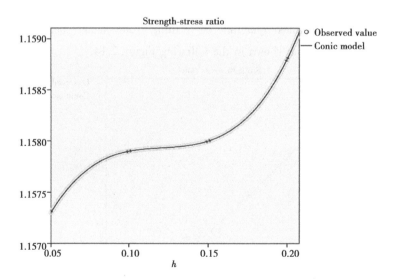

Figure 2.15 Cubic curve fitting diagram

$$\frac{\sigma_\theta}{\sigma_c} = 1.155 + 0.071R - 0.8R^2 + 1.6R^3 \qquad (2.17)$$

Seeing from the Figure 2.15, we find that with the increases of the concrete thickness, σ_θ/σ_c

also increases, i.e. the thicker concrete layer inhibits the unloading rebound of the surrounding rocks after excavation, which leads to an increase in σ_θ; thus in turn it leads to the corresponding increase of σ_c. However, the result is still smaller than the unsupported σ_θ/σ_c, thus confirming that some simple lining support has a certain inhibitory effect on rockburst activities; but as the supporting thickness increases, the suppression effect on the rockbursts slows down.

In order to further explore the impact of supporting strength on the intensity of rockbursts, we make some corresponding adjustments based on the above numerical model. After inputing the strength of sprayed concret into the numerical model, making simple anchor spray support in the full-face excavation, the corresponding horizontal and vertical stresses are obtained. After substituting them into the tangential principal stress formula, we calculate the corresponding stress-intensity ratios. The specific results are shown in the following Table 2.18.

Table 2.18 Strength ratio of sprayed concrete and stress-intensity ratio

σ_c	σ_t	E	ν	K	G	φ	C	Concrete grade	Strength-stress ratio
100.000	10.000	38.500	0.250	25.667	15.400	40.000	20.000	C20	1.1588
100.000	10.000	38.500	0.250	25.667	15.400	40.000	20.000	C25	1.1552
100.000	10.000	38.500	0.250	25.667	15.400	40.000	20.000	C30	1.1564
100.000	10.000	38.500	0.250	25.667	15.400	40.000	20.000	C35	1.1585
100.000	10.000	38.500	0.250	25.667	15.400	40.000	20.000	C40	1.1569

According to the thickness rules of sprayed concrete, the increase of concrete strength should also restrain the unloading rebound of the surrounding rocks after excavation, resulting in an increase in σ_θ, which in turn leads to a corresponding increase in σ_θ/σ_c. However, the simulation results show that σ_θ/σ_c fluctuates within a certain range. Therefore, it can be inferred that the intensities of different concrete grades have little difference. In those high-intensity-rockburst-prone areas, the grades of concrete have no obvious effect on the suppression of rockbursts.

The above analysis shows that in terms of excavation methods, stepwise excavation can reduce, to a certain extent, the intensity of rockbursts more effectively than full-face excavation; that the supporting construction does help to suppress rockburst activities. But the intensity of support is generally positively related to rockburst activities, suggesting that in practical tunnel projects, the flexible support should be applied as much as possible in order to reduce the intensity of rockbursts.

2.6 Recommended Ideas for Controlling High-intensity Rock Bursts

The above analysis shows that the intensity of rockbursts is related to the physical and mechanical properties of surrounding rocks, since they affect the energy storage of rocks; and that the stress release process of rockbursts is related to the excavation methods.

The root cause of a rockburst is that tunnel excavation makes the microelements in surrounding rocks change from the original three-way geostress state to the two-way geostress state, which strips the hard and brittle surrounding rocks from the original rock mass, especially in the high geostress area, which is also a horizontal structure developing area, where the rock chips ejecting directly from the surrounding rocks. According to the above numerical simulations results, based on the characteristics of rockbursts, we make the following recommendations to the treatment of high intensity rockbursts:

The advance drilling de-stress method: In the forward position of the tunnel, we can drill holes at a certain angle with the tunnel axial direction. Then the high stress in rocks can be released in advance, to reduce or even eliminate rock bursts. However, the amount of stress released in this way is so limited that its effect on high-intensity rockbursts is not obvious in real projects. Therefore, according to the above-mentioned numerical simulation results, we need to increase the diameter of the advance borehole as much as possible, so as to create a larger space for the redistribution of stress.

The controlled blasting pre-release method: we can disturb the surrounding rocks by huge energy released from blasting vibration and shock waves, to guide the redistribution of stress in the surrounding rocks, that is, by partially releasing the stress to reduce the rockburst intensity. Through optimizing the design of a series of factors, such as the pattern of blast holes, the diameter of blast holes, the depth of blast holes, the angle between the blast hole and the axial direction of the tunnel, the amount of charge, the method of charging, and the method of initiation, etc., we try to blast the tunnel in a way that the high geostress rocks in the tunnel are disturbed as violently as possible to releas high geostress without affecting the stability of the tunnel.

The inflatable anchor method: The inflatable anchor, also called the hydraulic expansion anchor, consists of the end sleeve, the retaining ring, the liquid injection end, the trays and other accessories. It is a double-layered tubular rod made of a seamless steel pipe, with an outer diameter greater than the hole diameter. After the pipe is placed in the anchor hole, it is expanded by high-pressure water. During the expansion process, the inflatable anchor rod is permanently deformed, making the wall of the anchor rod and the irregular wall of the anchor hole completely fitted. The friction and self-locking forces generated by the entire anchor rod further strengthen the rock mass. Thus the tunnel is strengthened to prevent rockburst damage. In real projects, when a low-intensity rockburst occurs, this method does have a certain preventive effect; but when a high-intensity rockburst is encountered, the anchor rod is often directly damaged, leading to no effect. So this method is not applicable to severe rock bursts.

The steel mesh reinforced with ultra-fine zeolite powder: The superfine zeolite powder allows concrete to harden quickly to strengthen the supporting effect. The steel mesh reinforcement made of sprayed concrete helps the surrounding rocks in tunnel as a whole to share the relatively high geostress. When the rockburst intensity is low, it has a good effect on suppressing rockburst; while when the rockburst intensity is high, the effect on suppressing rockburst is not very obvious.

The avoidance method: When a severe rockburst breaks out, avoidance may be the best way, withdrawing all personnel and equipments, waiting for operation after the rockburst. This is a passive and reluctant choice, used only when there is no other effective measures. Due to the suddenness and concealment of rock bursts, and the irregularity of time, it is difficult to avoid, so absolute safety cannot be guaranteed. What's more, avoidance might cause a substantial delay in the construction period or a waste of labor and equipment.

In summary, for minor and low-level rockbursts, those measures such as advance drilling, inflatable anchor, and spray of concrete can be used to control; for severe rockbursts, the above measures may have no obvious effects, so the passive avoidance method and the active pre-release de-stress blasting method are recommended. The huge energy, released by blasting vibration and shock waves, can disturb the surrounding rocks, guiding the redistribution and partial release of stress, and thus the intensity of rockburst can be reduced.

2.7 Summary

In this chapter, through the numerical simulation of rock's physical and mechanical properties and the excavation conditions, and the comparative analysis of rockburst prevention measures, the following conclusions are drawn:

①The main factor F_1, which is determined by the compressive strength and the internal friction angle, represents the responsiveness of rock intensity changes in rockburst activities, and its variance contribution rate to the variable has reached 76.889%, which indicates that F_1 is an important aspect in the response system of rockburst activities; the main factor F_2, determined by the elastic modulus, and the main factor F_3, mainly determined by cohesion, are also important to rockburst activities. The relationships among them are:

When F_2 and F_3 remain unchanged and only F_1 is changed, with the increase of F_1, the stress intensity ratio gradually decreases, which means that with the increase of F_1, rockburst activity gradually weakens.

$$\frac{\sigma_\theta}{\sigma_c} = 2.3945 - 10.8195 F_1 + 15.6245 F_1^2$$

When F_1 and F_3 remain unchanged and only F_2 is changed, with the increase of F_2, the stress intensity ratio also gradually increases, which means that with the increase of F_1, rockburst activity gradually increases.

$$\frac{\sigma_\theta}{\sigma_c} = 0.6881 + 0.1219 F_2^2 + 0.1581 F_2^3$$

When F_1 and F_2 remain unchanged and only F_3 is changed, with the increase of F_3, the stress intensity ratio gradually decreases, which means that with the increase of F_3, the rockburst activity gradually weakens.

$$\frac{\sigma_\theta}{\sigma_c} = 1.0031 - 0.799 F_3 + 0.2816 F_3^2$$

②The numerical simulations of the excavation conditions show that, compared with the full-face excavation, σ_θ/σ_c in stepwise excavation is reduced, which indicates that stepwise excavation is more helpful in reducing the intensity of rockburst than full-section excavation. Moreover, supporting helps to suppress rockburst activity. With the increase of the thickness of the support, σ_θ/σ_c also increases. It is the thicker concrete layer that inhibits the unloading rebound in the surrounding rocks after excavation, which results in an increase in σ_θ and then the corresponding increase of σ_θ/σ_c. However, this result is still smaller than the unsupported σ_θ/σ_c, which proves that the lining structure does connect the surrounding rocks as a unified whole to jointly bear the stress and thereby inhibiting rockburst activities to a certain extent.

③For the prevention and control of the severe rockbursts, the currently available active measure is the pre-release de-stress blasting method, where the huge energy released from blasting vibration or shock waves disturbs the surrounding rocks, guiding the redistribution and partial release of stress in the surrounding rocks, thereby reducing the intensity of rockbursts.

Chapter 3 Mechanism of the Pre-release De-stress Blasting for Rockbursts

3.1 Definition of De-stress Blasting

Pressure relief blasting first appeared in the 1950s. It is a method of releasing concentrated or high stresses around the tunnel through blasting, and then releasing them through explosive cracks or transferring them to deep rocks. The purpose is to reduce the impact of high-strength stress on the rock. It is an important artificial de-stress method in tunnels with high geostress. In the early 1950s, this method was first applied in gold mining in South Africa. The de-stress blasting, through the detonation effect of explosives, destroys rocks into cracks to redistribute the geostress in surrounding rocks, or changes the effective elastic modulus of rocks, so as to release high geostress or transfer them into deep surrounding rocks, and thus release geostress and improve the stress conditions in tunnels.

The de-stress blasting is an active preventing measure. On the one hand, the energy generated by the explosive directly acts on the rock mass, causing the rock mass to break and the physical structure to be destroyed, and making a large number of cracks in the rock, so that the elastic energy stored in rocks can be converted or consumed. On the other hand, the explosion in the blast hole forms a hollow surface, which can directly release the high geostress stored in the rock mass and form unloading cracks. These two factors interact with each other and work together to release the high stress in rock mass and transfer the peak stress in surrounding rocks in to deeper part; thus the intensity and energy of rock mass are greatly weakened, so as to reduce the probability of rock bursts. The purpose of de-stress blasting is to increase the range of fissure zone and generate more micro-facets, so that the stress is released to the facets, reducing the concentration of high stress, damaging the integrity of rock mass, and reducing the energy concentration of the rock mass. Since the rock fragmemtation degree in the crushing zone is extremely high, most of which are repeated crushing with high energy consumption, and the crushing area is the main area for de-stress blasting, in applying this method, the blasting parameters should be controlled properly to reduce the scope of crushing area, so that most of the energy can be applied to the expansion of the fractured area as much as possible, so as to achieve a better de-stress blasting effect.

The de-stress blasting can effectively redistribute the stress in surrounding rocks, or transfer the stress concentration area into deeper rocks. Figure 3.1 shows the changes of stress in surrounding rocks before and after the de-stress blasting in the tunnel. As applied in this figure, after the de-stress blasting, the stress peak has obviously shifted to the deeper part of surrounding rocks, thereby reducing the stress concentration in the surrounding rocks near the tunnel palm, and effectively changing the stress distribution in the surrounding rocks, so as to reduce the probability of rock bursts on the tunnel face, and protect the tunnel.

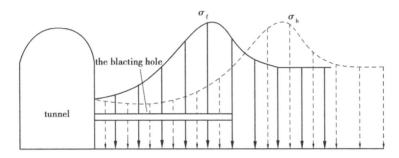

Figure 3.1   Changes in peak stress in surrounding rocks before and after tunnel de-stress blasting

3.2   Microscopic Mechanism of the De-stress Blasting

Blasting in rock mass is a technique which uses the heat and shock waves generated by explosives in the blasthole to break, loosen, and even throw rocks, to achieve the expected goal. The internal energy released by explosives in an instant and the expansion of gas generated by explosion work together on the surrounding medium, causing a huge damaging effect. In the blasting process, a crushing zone, a crack zone, and a vibration zone (disturbance zone) are sequentially formed around the blast hole, as shown in Figure 3.2. At the instant of explosion, a high temperature of several thousand degrees and a high pressure of tens of megapascals are generated, forming an explosive shock wave of several kilometers per second. The rocks, which are the closest to the explosion point, affected by the shock wave and the high-temperature high-pressure explosive gas, and generate some extreamly high radial and tangential stresses far exceeding the dynamic compressive

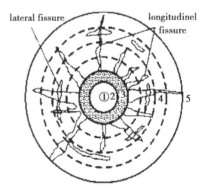

Figure 3.2   Internal blasting in rock mass media

1—the original blast hole   2—the enlarged cavity   3—the crushing area   4—the fissure area
5—the vibration area

intensity of the rocks, thus forming a compression crushing zone (crushing zone). The radius of the crushing zone can be estimated by the following formula (Dong Zhaoxing et al., 2005):

$$R_c = \left(\frac{\rho_m c_P^2}{5\sigma_c}\right)^{0.5} R_k \qquad R_k = \left(\frac{P_w}{\sigma_0}\right)^{0.25} r_b \tag{3.1}$$

Where R_c is crushing zone radius, R_k is limit value of cavity radius, σ_c is uniaxial compressive strength of rock, ρ_m is rock density, c_P is p-wave velocity of rocks, r_b is blast hole radius, P_w is average explosive pressure of explosives, and σ_0 is strength of rock under multidirectional stress. Its value is:

$$\sigma_0 = \sigma_c \left(\frac{\rho_m c_P}{\sigma_c}\right)$$

The initial value of the shock wave in the crushing area is high, but the decay is fast, so the radius of the crushing area should not be too large, which is often 2-5 times of the charging radius. With the rapid consumption of shock wave energy, the shock waves decay gradually into the compressive stress waves and continue to transmit radially in the rock. The radial stress waves transmit alternately between tension and compression. When the radial compressive stress value of the stress wave is lower than the compressive intensity of the rock, the rock will not undergo compressive failure, but only cause the radial displacement of the rock's particle. Because the rock receives tensile stress in the tangential direction while receiving radial compressive stress, and the rock is a brittle medium, whose tensile strength is low, so when the shear stress is greater than the tensile strength, the rock is broken and forms some radical crushes, which are interlinked with the crushed area. After the stress wave, the high-pressure explosive gas acts in the form of quasi-static pressure on the radial cracks in the cavity caused by the stress wave. Under the expansion, compression and wedge of the explosive gas, the crack continues to expand and extend, and the stress concentration caused by the gas at the crack tips also accelerates the expansion of the crack. This creates a cracked zone. Outside the cracked area, due to the continuing acts of the explosive shock waves and the blasting seismic waves upon the surrounding rocks (while it can not form a large number of fissures), the mechanical properties of the surrounding rocks are affected by the blasting disturbance, and even some cracks or micro-fissures are formed, which reduces the rock intensity, and thus this area turns to the vibration area or the disturbance area.

Affected by the strong compression of shock waves and stress waves, some elastic strain energy is accumulated in the rock. When the crushing zone is formed, the radial cracks are expanded, and the pressure of the explosive gas in the explosion chamber is reduced to a certain level; the previously accumulated energy will be released and converted into the unloading waves, which are transmitted to the explosion center, causing the particles to move concentrically. When these tensile stress waves are greater than the tensile intensity of the rock, the rock is broken and a hoop fracture is formed in the burst cavity. The staggered generation of radial fissures and annular fissures forms a fissured area, where the radial fissure plays a leading role outside the crushing area. Rock blasting mainly relies on the rupture zone formed by the crushing zone and the fracture zone. The radius of the rupture zone can be calculated as follows:

(1) Calculated by the stress wave of explosion

The radial crack is caused by tensile stress. The peak of tangential stress in the rock mass decreases with the distance to a certain rule. Replacing the peak value of tangential tensile stress $\sigma_{\theta max}$ with the tensile intensity of the rock, the radius of radial crack around the blast hole can be obtained:

$$\sigma_{\theta max} = \frac{bp_r}{\bar{r}^\alpha} \quad R_p = \left(\frac{bp_r}{\sigma_r}\right)^{\frac{1}{\alpha}} r_b \tag{3.2}$$

Where R_p is failure zone radius, p_r is hole wall initial impact pressure peak, σ_t is rock tensile intensity, α is stress wave decay coefficient, and r_b is blast hole radius.

(2) Calculated by quasi-static pressure of explosion gas

After the shock wave, the explosive gas expands in the medium-entropy of the blasthole. The pressure of the explosion gas which fills the blasthole is:

$$P_0 = \frac{1}{8}\rho_e D_e^2 \left(\frac{d_c}{d_b}\right)^6 \tag{3.3}$$

Where ρ_e is explosive density, D_e is explosive velocity, d_c is drug roll diameter, d_b is blast hole diameter.

According to the thick-walled cylinder model theory in elasto-plastic mechanics, the stress state in the rock is obtained, which shows the values of radial compressive stress are equal to the tangential tensile stress, that is:

$$\sigma_\theta = |\sigma_r| = \left(\frac{r_b}{r}\right)^2 p_0 \tag{3.4}$$

Where r is distance from the center of the blast hole, r_b is blast hole radius, σ_r is radial compressive stress, and σ_θ is tangential tensile stress.

Similarly, the tensile strength σ_t of the rock is used to replace the tangential tensile stress σ_θ in the above formula, then the radius of the fracture zone can be obtained as:

$$R_p = \left(\frac{p_0}{\sigma_t}\right)^{\frac{1}{2}} r_b \tag{3.5}$$

The areas outside the near explosion area (compressive crushing area) and the middle area (rupture area) are called far explosion areas, that is, the vibration areas. The stress wave in the vibration zone has been greatly attenuated, and it gradually tends to a periodic sine wave. At this time, the stress value can not cause damage to the rock, but cause the elastic vibration of the particle in the rock to form a seismic wave. The seismic wave traveled far enough until the explosion energy is completely absorbed by rocks. According to some scholars, the radius of the vibration zone can be estimated as follows:

$$R_s = (1.5 \text{ to } 2.8) \sqrt[3]{Q} \tag{3.6}$$

Where R_s is vibration zone radius, and Q is charge amount.

3.3 Derivation of the Calculation Formula of Residual Stress after De-stress Blasting

3.3.1 Derivation of de-stress blasting equation on two-dimensional plane

In transmiting into rocks, the blasting stress wave follows the exponential decaying law as follows (Ma Wenwei et al., 2015):

$$P = \sigma_t = P_2 \left(\frac{r}{r_b}\right)^{-\alpha} \tag{3.7}$$

Where σ_t is peak radial stress, r is distance from the center of the blast hole, r_b is blast hole radius, and α is stress wave decay coefficient.

Under the condition of cylindrical coupling charge, the pressure P_2 of the detonation wave generated by the explosive explosion into the rock is:

$$P_2 = P_1 \frac{2}{1 + \rho_0 D_v / \rho_r C_p} \tag{3.8}$$

Where ρ_0 is explosive density, D_v is explosive velocity, ρ_r is rock density, C_p is elastic wave velocity in rocks, and P_1 is the initial pressure of the detonation product, which can be calculated from the following formula:

$$P_1 = \frac{1}{4} \rho_0 D_v^2 \tag{3.9}$$

Under the condition of noncoupling charge, the impact pressure P_i on the blasthole rock wall is:

$$P_1 = \frac{1}{8} \rho_0 D_v^2 \left(\frac{d_c}{d_b}\right)^{2\alpha} \left(\frac{l_c}{l_b}\right)^{\alpha} n \tag{3.10}$$

Where d_c is blast hole diameter, d_b is charge diameter, l_c is blast hole length, l_b is charge length, n is pressure increase index, and α is pressure decay index.

Around the center of explosion, a shock wave transmits in the rock, and its decay index is:

$$\alpha = 2 + \frac{\mu}{1 - \mu}$$

Outside the shock wave area, a stress wave transmits in the rock, and its decay index is:

$$\alpha = 2 - \frac{\mu}{1-\mu}$$

The stress wave continues to decay and becomes a seismic wave. The decay rule of stress σ is:

$$\sigma = k\rho_r C_p \left(\frac{Q}{r}\right)^\alpha \tag{3.11}$$

Where μ is poisson's ratio, k is rock-related coefficients, Q is explosive quality of one detonation, and α is decay index, Generally 1-2.

From Equation $\sigma_0 = \sigma_c \left(\frac{\rho_m C_p}{\sigma_c}\right)^{0.25}$, we obtain:

$$\sigma_0 = \sigma_c \left(\frac{\rho_m c_p}{\sigma_c}\right)^{0.25} = \left(\frac{\sigma_c^4 \rho_m c_p}{\sigma_c}\right)^{0.25} = (\sigma_c^3 \rho_m c_p)^{0.25} \tag{3.12}$$

Substituting the above formula into formula (3.1), we acquire:

$$R_c = \left(\frac{\rho_m c_p^2}{5\sigma_c}\right)^{0.5} \left(\frac{P_w}{\sigma_c^{0.75} \rho_m^{0.25} c_p^{0.25}}\right)^{0.25} r_b = \left(\frac{\rho_m^{1.75} c_p^{3.75} P_w}{25\sigma_c^{2.75}}\right)^{0.25} r_b \tag{3.13}$$

According to formula (3.7), taking any point on the working face (in the very close range, that is, the distance is less than 2-5 times the blasthole radius), after the explosion, the explosive force transmitted to this point is:

$$p = p_2 \left(\frac{r_1}{r_b}\right)^{-\alpha}, \ p = p_2 \left(\frac{r_2}{r_b}\right)^{-\alpha}, \ p = p_2 \left(\frac{r_n}{r_b}\right)^{-\alpha}$$

The above formulas are superimposed:

$$P_t = P_2 r_b^\alpha (r_1^{-\alpha} + r_2^{-\alpha} + \cdots + r_n^{-\alpha}) \tag{3.14}$$

Near the center of explosion (the compressive crushing zone), i.e. $r \leq R_c = \left(\frac{\rho_m^{1.75} c_p^{3.75} P_w}{25\sigma_c^{2.75}}\right)^{0.25} r_b$,

The shock wave transmits in the rocks, and its decay index is:

$$\alpha = 2 + \frac{\mu}{1-\mu}$$

Outside the shock wave area (the rupture zone), i.e. :

$$\left(\frac{\rho_m^{1.75} c_p^{3.75} P_w}{25\sigma_c^{2.75}}\right)^{0.25} r_b = R_c \leq r \leq R_p = \left(\frac{bp_r}{\sigma_t}\right)^{\frac{1}{\alpha}} r_b$$

The stress wave transmits in the rocks, and the decay index given by scholars is:

$$\alpha = 2 - \frac{\mu}{1-\mu}$$

The stress wave continues to decay and becomes a seismic wave (in the vibration zone). From formulas (3.3) and (3.5), the radius of the vibration zone is:

$$r \geqslant R_{\mathrm{p}} = \left(\frac{bp_{\mathrm{r}}}{\sigma_{\mathrm{t}}}\right)^{\frac{1}{\alpha}} r_{\mathrm{b}}$$

Then we can get $R_{\mathrm{p}} = (2-5) r_{\mathrm{b}}$, which means when the distance of any point to the center of the blasthole is greater than 2–5 times the radius of the blast hole, the effect of blasting on rocks mainly depends on the seismic waves, and its decay rule obeys formula (3.11). From (3.11), the decay coefficient of the stress wave, transmitted to a certain point in the vibration zone can be obtained as:

$$\sigma_{\mathrm{t}} = k\rho_{\mathrm{r}} C_{\mathrm{p}} Q^{\alpha} (r_1^{-\alpha} + r_2^{-\alpha} + \cdots + r_n^{-\alpha})$$

Assuming that the blast holes are arranged in a semi-circular shape with a total number of m, the semi-circular arc is divided into $m-1$ segments by the blast eyes, as shown in Figure 3.3. The n-th blasthole is A: (x_0, y_0), and any de-stress point B is taken, and its coordinates are (x, y). Then the angle between point A and coordinate origin O is:

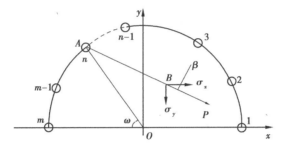

Figure 3.3 Layout of de-stress blast holes

P—Blasting stress σ_x—Blasting horizontal stress component

σ_y—Blasting vertical stress component

$$\omega = \pi \frac{n-1}{m-1}$$

Let the radius of the semi-arc be R, we have:

$$x_0 = R \cos \alpha, \ y_0 = R \sin \alpha$$

$$\cos \beta = \frac{|x - x_0|}{\sqrt{(x - x_0)^2 + (y - y_0)^2}} = \frac{|x - R \cos \alpha|}{\sqrt{(x - R \cos \alpha)^2 + (y - R \sin \alpha)^2}}$$

$$\sin \beta = \frac{|y - y_0|}{\sqrt{(x - x_0)^2 + (y - y_0)^2}} = \frac{|y - R \sin \alpha|}{\sqrt{(x - R \cos \alpha)^2 + (y - R \sin \alpha)^2}}$$

by: $\sigma_x = P \cos \beta$, $\sigma_y = P \sin \beta$. According to the above formula:

$$\sigma_{Bx} = P \cos \beta = p \frac{|x - x_0|}{\sqrt{(x - x_0)^2 + (y - y_0)^2}} = P \frac{|x - R \cos \alpha|}{\sqrt{(x - R \cos \alpha)^2 + (y - R \sin \alpha)^2}}$$

$$= p \frac{\left|x - R\cos\frac{\pi(n-1)}{m-1}\right|}{\sqrt{\left[x - R\cos\frac{\pi(n-1)}{m-1}\right]^2 + \left[y - R\sin\frac{\pi(n-1)}{m-1}\right]^2}}$$

$$\sigma_{By} = P\sin\beta = p\frac{|y - y_0|}{\sqrt{(x-x_0)^2 + (y-y_0)^2}} = P\frac{|y - R\sin\alpha|}{\sqrt{(x - R\cos\alpha)^2 + (y - R\sin\alpha)^2}}$$

$$= p\frac{\left|y - R\sin\frac{\pi(n-1)}{m-1}\right|}{\sqrt{\left[x - R\cos\frac{\pi(n-1)}{m-1}\right]^2 + \left[y - R\sin\frac{\pi(n-1)}{m-1}\right]^2}}$$

$$\sigma_{Bxt} = P_1\cos\beta_1 + P_2\cos\beta_2 + \cdots + P_n\cos\beta_n$$

$$= P_1\frac{|x - x_{01}|}{\sqrt{(x - x_{01})^2 + (y - y_{01})^2}} + P_2\frac{|x - x_{02}|}{\sqrt{(x - x_{02})^2 + (y - y_{02})^2}} + \cdots + P_n\frac{|x - x_{0n}|}{\sqrt{(x - x_{0n})^2 + (y - y_{0n})^2}}$$

$$= P\left(\frac{r_1}{r_b}\right)^{-\alpha}\frac{|x - x_{01}|}{\sqrt{(x - x_{01})^2 + (y - y_{01})^2}} + P\left(\frac{r_2}{r_b}\right)^{-\alpha}\frac{|x - x_{02}|}{\sqrt{(x - x_{02})^2 + (y - y_{02})^2}} + \cdots + P\left(\frac{r_n}{r_b}\right)^{-\alpha}\frac{|x - x_{0n}|}{\sqrt{(x - x_{0n})^2 + (y - y_{0n})^2}}$$

$$= Pr_b^\alpha\frac{|x - x_{01}|}{[(x - x_{01})^2 + (y - y_{01})^2]^{\frac{\alpha+1}{2}}} + Pr_b^\alpha\frac{|x - x_{02}|}{[(x - x_{02})^2 + (y - y_{02})^2]^{\frac{\alpha+1}{2}}} + \cdots + Pr_b^\alpha\frac{|x - x_{0n}|}{[(x - x_{0n})^2 + (y - y_{0n})^2]^{\frac{\alpha+1}{2}}}$$

Where r is distance between initiation point and de-stress point, r_b is blast hole radius. Similarly:

$$\sigma_{Byt} = Pr_b^\alpha\frac{|y - y_{01}|}{[(x - x_{01})^2 + (y - y_{01})^2]^{\frac{\alpha+1}{2}}} + Pr_b^\alpha\frac{|y - y_{02}|}{[(x - x_{02})^2 + (y - y_{02})^2]^{\frac{\alpha+1}{2}}} + \cdots + Pr_b^\alpha\frac{|y - y_{0n}|}{[(x - x_{0n})^2 + (y - y_{0n})^2]^{\frac{\alpha+1}{2}}}$$

$$x_0 = R\cos\alpha, \quad y_0 = R\sin\alpha, \quad \alpha = \frac{\pi(n-1)}{m-1}$$

Therefore, after blasting, the explosive force at any point in the vicinity of the working face is:

$$\begin{cases} \sigma_{xt} = Pr_b^\alpha \sum_{n=1}^{m} \dfrac{\left| x - R \cos \dfrac{\pi(n-1)}{m-1} \right|}{\left\{ \left[x - R \cos \dfrac{\pi(n-1)}{m-1} \right]^2 + \left[y - R \sin \dfrac{\pi(n-1)}{m-1} \right]^2 \right\}^{\frac{\alpha+1}{2}}} \\ B_{yt} = Pr_b^\alpha \sum_{n=1}^{m} \dfrac{\left| x - R \sin \dfrac{\pi(n-1)}{m-1} \right|}{\left\{ \left[x - R \cos \dfrac{\pi(n-1)}{m-1} \right]^2 + \left[y - R \sin \dfrac{\pi(n-1)}{m-1} \right]^2 \right\}^{\frac{\alpha+1}{2}}} \end{cases} \quad (3.15)$$

Because within the extremely close zone, the blasting stress is much greater than the intensity of the rock, which results in a rock failure, and the residual stress is close to 0. In addition, since the range of the extremely close zone is usually small, which is generally 2-5 times the blasthole radius, the de-stress blast hole is usually arranged outside the extremely close zone. Therefore, the residual stress of de-stress blasting should focus on the effects of seismic waves outside the extremely close zone. The blasting stress on any point of the surrounding rocks in the faraway area is:

$$\begin{cases} \sigma_{Vr} = \sigma_V - k\rho_r C_p Q^\alpha (r_1^{-\alpha} \cos \beta_1 + r_2^{-\alpha} \cos \beta_2 + \cdots + r_n^{-\alpha} \cos \beta_n) \\ \sigma_{Hr} = \sigma_H - k\rho_r C_p Q^\alpha (r_1^{-\alpha} \sin \beta_1 + r_2^{-\alpha} \sin \beta_2 + \cdots + r_n^{-\alpha} \sin \beta_n) \end{cases} \quad (3.16)$$

Where σ_{Vr} is x-axis residual stress value at de-stress point, σ_{Hr} is y-axis residual stress value at de-stress point, σ_V is x axial geostress, σ_H is y axial geostress, $r_1, r_2, \cdots, r_n$ are the distances between the de-stress point and each borehole, among them: $r_n = \sqrt{(x-x_n)^2 + (y-y_n)^2}$, and β_1, $\beta_2, \cdots, \beta_n$ are the angle of blasting stress of each blasthole with the x axis at the de-stress point, among them:

$$\begin{cases} \cos \beta_n = \dfrac{|x - x_n|}{\sqrt{(x-x_n)^2 + (y-y_n)^2}} \\ \sin \beta_n = \dfrac{|y - y_n|}{\sqrt{(x-x_n)^2 + (y-y_n)^2}} \end{cases}$$

Where K is rock-related coefficient, Q is charge, α is stress decay index, Generally 1-2. $\alpha = 2 - \dfrac{\mu}{1-\mu}$, ρ_r is rock density, and C_p is velocity of elastic waves in rocks.

Generally, the range of the blasting close zone is very small, only 2 to 5 times the blast hole radius. After blasting, the surrounding rocks are in a completely broken or fissured state, which can be considered as close to a complete de-stress. Thus it is of little significance to conduct a theoretical deduction, and we'll not take into consideration the residual stress equation of de-stress blasting in the close area in later derivation process, instead, we'll only discuss the residual stress equation of de-stress blasting in those far areas.

3.3.2 Deduction of the de-stress blasting equation on 3-dimensional plane

Let the coordinates of the de-stress point B be (x, y, z), and the angles between the line (where the stress wave is transmitted to point B) and the X, Y, Z axes are δ, β, γ, Then there is: $\cos^2\delta + \cos^2\beta + \cos^2\gamma = 1$. Let the n-th blast hole be A, and the inclination angle of the blast hole along the tunnel axis is φ, the coordinates are (x_0, y_0, z_0), then: $x_0 = R\cos\omega$, $y_0 = R\sin\omega$, $\omega = \pi\dfrac{n-1}{m-1}$, among which m is the number of blastholes.

According to the trigonometric function relationship, $\cos\delta$, $\cos\beta$, $\cos\gamma$ can be represented as follows:

$$\cos\delta = \frac{|x - x_0|}{\sqrt{(x - x_0)^2 + (y - y_0)^2 + (z - z_0)^2}},$$
$$\cos\beta = \frac{|y - y_0|}{\sqrt{(x - x_0)^2 + (y - y_0)^2 + (z - z_0)^2}}, \qquad (3.17)$$
$$\cos\gamma = \frac{|(z - z_0)|}{\sqrt{(x - x_0)^2 + (y - y_0)^2 + (z - z_0)^2}},$$

$$\sigma_x = P\cos\delta \quad \sigma_y = P\cos\beta \quad \sigma_z = P\cos\gamma$$

Similarly, the total de-stress blasting stress at any point is:

$$\begin{cases}\sigma_{xt} = k\rho_r C_p Q^\alpha (r_1^{-\alpha}\cos\delta_1 + r_2^{-\alpha}\cos\delta_2 + \cdots + r_n^{-\alpha}\cos\delta_n) \\ \sigma_{yt} = k\rho_r C_p Q^\alpha (r_1^{-\alpha}\cos\beta_1 + r_2^{-\alpha}\cos\beta_2 + \cdots + r_n^{-\alpha}\cos\beta_n) \\ \sigma_{zt} = k\rho_r C_p Q^\alpha (r_1^{-\alpha}\cos r_1 + r_2^{-\alpha}\cos r_2 + \cdots + r_n^{-\alpha}\cos r_n)\end{cases} \qquad (3.18)$$

For the purpose of the de-stress blasting, the calculation method of the residual stress is: to use the horizontal geostress to minus the total horizontal blasting stress, which is: $\sigma_{Vr} = \sigma_V - \sigma_x$. Then there is:

$$\sigma_{Vr} = \sigma_V - k\rho_r C_p Q^\alpha (r_1^{-\alpha}\cos\delta_1 + r_2^{-\alpha}\cos\delta_2 + \cdots + r_n^{-\alpha}\cos\delta_n) \qquad (3.19)$$

Similarly, on the axis of the tunnel there is: $\sigma_{zr} = \sigma_z - \sigma_z$ through which the residual stress equation of de-stress blasting can be obtained as follows in 3-20:

$$\sigma_{zr} = \sigma_z - k\rho_r C_p Q^\alpha (r_1^{-\alpha}\cos\beta_1 + r_2^{-\alpha}\cos\beta_2 + \cdots + r_n^{-\alpha}\cos\beta_n) \qquad (3.20)$$

Vertically, calculating $\sigma_{Hr} = \sigma_H - \sigma_y$, we can obtain:

$$\sigma_{Hr} = \sigma_H - k\rho_r C_p Q^\alpha (r_1^{-\alpha}\cos\gamma_1 + r_2^{-\alpha}\cos\gamma_2 + \cdots + r_n^{-\alpha}\cos\gamma_n) \qquad (3.21)$$

In summary, the residual stress equations for de-stress blasting under three-dimensional conditions are:

$$\begin{cases}\sigma_{Vr} = \sigma_V - k\rho_r C_p Q^\alpha(r_1^{-\alpha}\cos\delta_1 + r_2^{-\alpha}\cos\delta_2 + \cdots + r_n^{-\alpha}\cos\delta_n) \\ \sigma_{zr} = \sigma_z - k\rho_r C_p Q^\alpha(r_1^{-\alpha}\cos\beta_1 + r_2^{-\alpha}\cos\beta_2 + \cdots + r_n^{-\alpha}\cos\beta_n) \\ \sigma_{Hr} = \sigma_H - k\rho_r C_p Q^\alpha(r_1^{-\alpha}\cos\gamma_1 + r_2^{-\alpha}\cos\gamma_2 + \cdots + r_n^{-\alpha}\cos\gamma_n)\end{cases} \quad (3.22)$$

Where σ_{Vr} is X-axis residual stress value at de-stress point, σ_{zr} is Z-axis residual stress value at de-stress point, σ_{Hr} is Y-axis residual stress at de-stress point, α is stress decay index, Generally $1-2. \alpha = 2 - \dfrac{\mu}{1-\mu}$, C_p is elastic wave velocity in rocks, σ_V is X axial geostress value, σ_z is Z axial geostress value, σ_H is Y axial geostress value, $r_1, r_2, \cdots, r_n$ are the distance between the de-stress point and each borehole, among them,

$$r_n = \sqrt{(x-x_n)^2 + (y-y_n)^2 + (z-z_n)^2}$$

$\delta_1, \delta_2, \delta_n, \beta_1, \beta_2, \beta_n, \gamma_1, \gamma_2, \gamma_n$ are the angle between the blasting stress of each blasthole and the x-axis, y-axis and z-axis at the de-stress point, among them,

$$\begin{cases}\cos\delta_n = \dfrac{|x-x_n|}{\sqrt{(x-x_n)^2+(y-y_n)^2+(z-z_n)^2}} \\ \cos\beta_n = \dfrac{|y-y_n|}{\sqrt{(x-x_n)^2+(y-y_n)^2+(z-z_n)^2}} \\ \cos\gamma_n = \dfrac{|z-z_n|}{\sqrt{(x-x_n)^2+(y-y_n)^2+(z-z_n)^2}}\end{cases}$$

ρ_r is rock density, k is coefficients related to rock properties and blasting parameters, and Q is charge amount.

In summary, by calculation, we can obtain the residual stress value at any point of the tunnel after the interaction between the detonation force and the geostress in blasting, and based on the rockburst occurrence criterion (Equation 2.7), we can predict whether the rockburst is likely to occur, and achieve the rockburst prediction purpose. Then the optimal blasting parameters can be obtained by adjusting the relative coordinates of the blasthole and the de-stress point in the formula, the coefficient k related to rock properties and blasting parameters, and the charge amount Q, etc. through which we can control the energy generated after blasting to achieve the optimal de-stress effect and decrease the rockburst intensity. In addition, adjusting the diameter of the blasthole can also adjust the range of the extreamly close zone, but the adjustable range is so small that it is not very significant in engineering practices.

3.4 Role of the Stress Pre-release Loose Circle

Under the impact of blasting and other construction operations, an area is created in the rock mass around the tunnel. In this area, some internal cracks develope and the rock mass become loose, showing strong nonlinear and discontinuous features. The equilibrium stress in original rock is destroyed and the stress is redistributed: the radial stress is reduced, with stress near the surface close to zero, and the tangential stress is increased to form the stress concentration. This stress state of surrounding rocks changes from 3-dimension to 2-dimension. This area is called the stress pre-release loose circle.

The range of the stress loose circle can be measured on site. The on-site measurement mainly includes the sonic wave detection method, the multi-point displacement meter method, the geological radar method, the seismic wave method, the resistivity method, and the permeability method, etc. The sonic wave detection method is based on the transmission rule of elastic waves in the rock mass. The elastic wave's velocity, besides closely related to the physical prosperity of rocks, is also closely related to the stress state and the structure of the rock mass. Generally speaking, the elastic waves transmit faster in those hard rocks, in the rock mass with undeveloped fissures and low weathering, in the rock masses with small porosity, high density, and high elastic modulus, in the rock mass with high compressive strength, in the rock mass with few faults and fracture zones, or in the small-scale rock mass. Conversely, the elastic waves transmit more slowly.

Based on this feature of rock mass, the sonic wave detection method is used to determine the range of the loose circle. The steps are as follows:

①To determine the basic wave velocity of various surrounding rocks before excavation;

②To determine the criteria of the loose circle: According to the national standard (by the Industry Standards Compilation Group of People's Republic of China, 1999), the wave velocity method is used to judge the failure standards of the quality of blasting rock mass. The rate of changes is based on the wave velocity C_{p1} before the blasting and the wave velocity C_{p2} after the blasting, that is, according to the formula $n = 1 - C_{p1}/C_{p1}$, if $n > 10\%$, it can be judged as a loose circle;

③According to the sonic wave detection curve and the above-mentioned criteria, we can divide the loosen circles after blasting excavation in different parts of the tunnel, and classify them according to the rate of changes, and determine the thickness of each degree of loose circle;

④The same-level loose circles are connected into a closed area, so that we can obtain a hierarchical model for the loose circle classification.

Based on the characteristics of the surrounding rocks in the deep well, such as the elastoplastic state, the strain softening state and the residual strength state, Jing Hongwen et al. analyzed the infinite long circular tunnels under the state of hydrostatic stress. Through assuming that the surrounding rocks approximately meet the basic assumption of the elastoplastic theory and the weight of rocks is ignored, they obtained the calculation formula of the loose circle radius as following (Jing Hongwen et al., 2005):

$$R_p = r_0 \left\{ \frac{C_0[P_0 + \sigma_c/K_1 + (K_2 nP')/(K_1 K_3)]}{P_i + \sigma_c^*/K_1} - \frac{[2nP' + (1+\xi)(\sigma_c - \sigma_c^*)]/K_1 K_3}{P_i + \sigma_c^*/K_1} \right\}^{\frac{1}{K_1}}$$

Where r_0 is tunneling radius (m), P_0 is original rock stress in hydrostatic stress field (MPa), P_i is support resistance per unit area (Usually taken as 0 MPa), σ_c is ultimate unidirectional compressive strength of rock mass (MPa), σ_c^* is residual unidirectional compressive strength of rock mass (MPa), and ξ is parameters related to rock swelling; from the plastic increment theory, we get:

$$\xi = \frac{1 + \varphi'}{1 - \varphi'}$$

Among them $\varphi' = (0.7 \text{ to } 0.9)\varphi$, φ is the internal friction angle;

$K_1 = K_p - 1$, $K_2 = K_p + 1$, $K_3 = K_p + \xi$, $K_p = (1 + \sin \varphi)/(1 - \sin \varphi)$

n is the softening coefficient of rock, $n = M_0/E$, among them E is the elastic modulus of rock, M_0 is the strain damping modulus ($M_0 = \tan \alpha$);

$P' = 0.5(1 + \mu)[2P_0 \sin \varphi + (1 - \sin \varphi)\sigma_c]$ (μ is poisson's ratio):

$$C_0 = \frac{2}{K_2} \left[\frac{2nP'}{2nP' + (1+\xi)(\sigma_c - \sigma_c^*)} \right]^{\frac{K_1}{1+\xi}}$$

Therefore, the thickness of the loose circle (the rupture zone) is: $L_p = R_p - r_0$.

There are many factors affecting the loose circles, but they are mainly determined by the original rock stress P_0. The original rock stress field is dominated by gravity or tectonic stress. Some scholars have found that when the intensity of rock keeps constant, the loose circle will become thicker with the increase of the original rock stress (Jing Hongwen et al., 1999). Through on-site tests, Jing Hongwen et al. simulated the test results and concluded that the relationships among the original rock stress P_0, the uniaxial compressive strength σ_c of the rock, and the loose circle thickness L_p are as follows:

$$L_p = 1.293(P_0/\sigma_c - 0.379)^{1/2}$$

3.5 Factors Influencing De-stress Blasting Effectiveness

De-stress blasting is an effective method to reduce high geostress and eliminate rock bursts. Actually, there are many influencing factors, generally including: the rock properties, charge amounts, explosive properties, blasthole parameters, and the initiation methods.

The physical and mechanical properties of rocks are important factors affecting the occurrence of rock bursts, and affecting the effect of de-stress blasting. Generally speaking, such parameters as the compressive strength and the elastic modulus of rocks have a significant effect on rockbursts, which has been discussed in Chapter 2, i.e., the main factor F_1, determined by the two indicators of the compressive strength and the internal friction angle, represents the responsiveness of rock's intensity changes to rockburst activities, and thus it is an important factor in the response system of rockburst activities; the changes of principal factor F_2, determined by the elastic modulus, and the changes of the main factor F_3, determined by cohesion, are also important to the rockburst activities.

From the perspective of blasting, since the brittle rocks are more easily broken by stress waves and expansion of explosive gas, etc., the brittle rocks with moderate hardness can achieve better de-stress blasting effects. Therefore, the intensity of rocks in lithology, especially the tensile and compressive intensity, has a significant effect on de-stress effects.

The amount of charge determines the force to the broken rocks and decides the blasting and crushing scope. Different charge amounts produce different de-stress effects. The larger the charge amount, the weaker the stress concentration of the surrounding rocks in front of the working face; the lower the peak stress, the larger the range of damage to the rock mass, and the wider the stresse-release zone. With the increase of charge amounts, although the reduction of the stress peak is not obvious, the peak stress area is obviously transferred into the deep rocks, which indicates that de-stress blasting cannot completely eliminate stress concentration; but it obviously push the peak stress area deeper, so as to greatly reduce the risk of rockbursts in the de-stress zone, and the charge amount is linearly related to the distance of stress transfer.

The distance between the stress-release point and the borehole significantly affects the effects of de-stress blasting. As shown in 3.2 of this chapter, when the distance is less than the crushing radius (usually 2 to 5 times the blasthole diameter, and this range is also called the extreamly close zone), the rock may be broken or cracked by the blasting stress waves and the expansion of the explosive gas, so that the geostress is released completely. When the stress-release point is located outside the extreamly close zone, the rock is mainly disturbed by the blasting seismic wave and its stress is changed. Since the seismic wave decays quickly, the distance is still the most prominent influencing factor.

In terms of blasting parameters, the properties of explosives, the blasting parameters, and the initiation methods restrict the performance of explosives. The rationality and scientificity of the arrangement of these factors are more conducive to the release effects. Thus to properly increase the charge density and to select those high detonation speed explosives, especially the explosives which can generate a large volume of gas during explosion, do lead to a large explosive cavity, and do improve the de-stress effect. The plane distance, the row distance, and the three-dimensional inclination of the blast hole will cause changes with de-stress effects. Different charging methods and initiation methods will also affect the de-stress effects. The aqueous medium uncoupled charging is suitable for the de-stress blasting under high-stress concentration, while the reverse initiation method can prolong the action time of stress waves and explosive gases, which can generate more cracks and further expand and extend the cracks. Under different buried depths, the de-stress blasting can push the stress concentration area deeper in different degrees, and the traversing distance gradually decreases as the buried depth increases.

3.6 Summary

This chapter analyzes the microscopic mechanism of de-stress blasting. De-stress blasting relieves the stress in rock by damaging the rock mass, producing cracks and absorbing blasting energy. By controlling the blasting parameters to affect the de-stress results, meanwhile by controlling the scope of the loose circle, we can reduce the stress concentration in rock mass or transfer it into the deeper surrounding rocks; thus we can achieve the effect of the control method on pre-release de-stress blasting, and ensure the stability of the loose circle after excavation.

By theoretically deriving the calculation formulas of the residual stress for de-stress blasting, we can calculate the residual stress value after the interaction between the detonation force and the geostress at any point outside the extremely close zone. In the extremely close zone, the blasting stress is much greater than the intensity of the rocks; thus rock failure occurs, at this time the residual stress in the rock after blasting is close to zero. In addition, because the range of the extremely close zone is very small, (usually 2 to 5 times the blast hole radius), the de-stress blastholes are usually arranged outside the extremely close zone of the blast hole. Therefore, it is meaningless to derive the residual stress formula for de-stress blasting in the extremely close zone.

The derived residual stress equations for de-stress blasting under three-dimensional conditions are:

$$\begin{cases} \sigma_{Vr} = \sigma_V - k\rho_r C_p Q^\alpha (r_1^{-\alpha}\cos\delta_1 + r_2^{-\alpha}\cos\delta_2 + \cdots + r_n^{-\alpha}\cos\delta_n) \\ \sigma_{zr} = \sigma_z - k\rho_r C_p Q^\alpha (r_1^{-\alpha}\cos\beta_1 + r_2^{-\alpha}\cos\beta_2 + \cdots + r_n^{-\alpha}\cos\beta_n) \\ \sigma_{Hr} = \sigma_H - k\rho_r C_p Q^\alpha (r_1^{-\alpha}\cos\gamma_1 + r_2^{-\alpha}\cos\gamma_2 + \cdots + r_n^{-\alpha}\cos\gamma_n) \end{cases} \quad (3.23)$$

Where σ_{Vr} is X-axis residual stress value at de-stress point, σ_{zr} is Z-axis residual stress value at de-stress point, σ_{Hr} is Y-axis residual stress at de-stress point, α is stress decay index, Generally $1-2.\alpha = 2-\dfrac{\mu}{1-\mu}$, C_p is elastic wave velocity in rock, σ_V is X axial geostress, σ_z is Z axial geostress, σ_H is Y axial stress value, and $r_1, r_2, \cdots, r_n$ are the distance between the de-stress point and each borehole, among them,

$$r_n = \sqrt{(x-x_n)^2 + (y-y_n)^2 + (z-z_n)^2}$$

$\delta_1, \delta_2, \delta_n, \beta_1, \beta_2, \beta_n, \gamma_1, \gamma_2, \gamma_n$ are the angle between the blasting stress of each blasthole and the X-axis, Y-axis and Z-axis at the de-stress point, among them,

$$\begin{cases} \cos \delta_n = \dfrac{|x - x_n|}{\sqrt{(x - x_n)^2 + (y - y_n)^2 + (z - z_n)^2}} \\ \cos \beta_n = \dfrac{|y - y_n|}{\sqrt{(x - x_n)^2 + (y - y_n)^2 + (z - z_n)^2}} \\ \cos \gamma_n = \dfrac{|z - z_n|}{\sqrt{(x - x_n)^2 + (y - y_n)^2 + (z - z_n)^2}} \end{cases}$$

ρ_r is rock density, k is coefficients related to rock properties and blasting parameters, and Q is charge amount.

De-stress blasting is an effective method to reduce high geostress and to eliminate rockbursts. According to the derived formulas, the residual stress is closely related to these factors: the distances and the relative coordinates between the blast hole and the de-stress test point, the charge amounts, the parameters of rock property, and the blasting parameters (including the properties of the explosives, the blast hole parameters, and the initiation method).

Chapter 4 Experimental Study on Rock Mechanics of Surrounding Rocks in Tunnels

In order to obtain the physical and mechanical parameters of the surrounding rocks in tunnels, laboratory tests or in-situ field tests need to be performed. The field test mainly studies the structure, intensity, and deformation characteristics of the rock mass, and the laboratory test mainly studies the microstructure, intensity, and deformation characteristics of the rock mass. After sampling on site, we also conduct some laboratory tests, for that the laboratory test is not restricted by field conditions, it is easy to simulate the stress conditions of rocks. In addition, a large number of rock samples are collected for testing, and through statistical analysis of those test data, the physical mechanics parameters of rocks have been obtained. Therefore, a more effective method is to firstly carry out some normal tests in the laboratory to obtain some necessary mechanical parameters, and then calculate them to get the intensity parameters of the rock mass. In this book, we firstly, through sampling laboratory tests, obtain some conventional index parameters, such as the compressive intensity and the elastic modulus of surrounding rocks. Then based on the project of Erlangshan Tunnel of Yakang Expressway, we establish some numerical simulation calculation models, with an aim to optimize a set of parameter combinations for de-stress blasting. Because the Erlangshan Tunnel has been open to traffic before the field test, there is no condition for verification tests. So we choose the Sangzhuling tunnel of the Linla Railway to carry out the field verification tests, since there are more prominent geostress phenomena in this tunnel. Therefore, the experimental samples mainly come from the Erlangshan Tunnel and the Sangzhuling tunnel. The uniaxial compression deformation tests, the conventional triaxial compression tests and the tensile strength tests are done on the sample rocks and the deformation parameters of the rocks are also calculated (Wei Mingyao et al., 2011).

4.1 Lab Environment

①Sampling. We take samples from the Andesite in the K76 + 054 to K76 + 280 sections of the Erlangshan Tunnel, and the gneiss in the K186 + 94.020 sections of the Sangzhuling tunnel, by renting drilling and coring equipment on site. The diameter of each core sample is no less than the minimum size of the test piece. After the samples are taken, they are stably transported back to the lab in the university. There are at least 5 groups of samples in each of the two tunnels, and at least 3 pieces in each group. The Rock Mechanics Laboratory in Chengdu University of Technology can process those samples to meet the needs of the experiment, including drilling, cutting, smoothing, and air-drying. After air-drying, the basic data, such as the geometric dimensions and the weights, are measured.

②Production of test equipments and test pieces. The test equipment and materials mainly are hydraulic programmable servo rigidity testing machine, static resistance strain gauge, resistance strain gauge, acetone, and some cements, etc.

Figure 4.1 Rock sample

The production method is to drill the rock block into a cylindrical core on a drilling machine to cut it into a cylinder with a cutting machine, and to grind it on a grinder into the specific size required for the experiment. The shape and dimension of the test piece are shown in Figure 4.1. The diameter of the sample is 50 mm, and the height-diameter ratio of the compression deformation test is 2∶1 (the height is 100 mm and the diameter is 50 mm). The height-diameter ratio of the sample for the the tensile test is about 1∶1 (50 mm in height and 50 mm in diameter). Cleaning the samples, air-drying them in natural state, and numbering them after air drying. The numbers of rock sample in Erlangshan Tunnel are: E1-1—E1-3、E2-1—E2-3…E5-1—E5-3. The numbers of rock sample in Sangzhuling tunnel are:S1-1—S1-3、S2-1—S2-3…S5-1—S5-3.

4.2 Test Methods and Results

In order to experimentally measure the mechanical parameters of rocks, such as the intensity, the elastic modulus and the ratio of poisson, to get the mechanical property curves of rocks, we mainly carried out the following test methods: the uniaxial tensile test (the Brazilian splitting test), the uniaxial compression deformation test, and the triaxial compression deformation test, et al..

4.2.1 Uniaxial compressive strength test

The main purpose of the uniaxial compression test on rocks is to measure the longitudinal strain and the transverse strain of rocks under the uniaxial stress, and draw the relation curves between the stress, and the longitudinal strain and the transverse strain. Then based on the features of the curve, we can determine the deformation characteristics of rocks. The parameters of rock deformation are mainly the elastic modulus and the ratio of poisson.

1) Experimental process

The displacement control method is selected for the test, and as to the loading rate, recommended by the International Professional Committee on Rock Mechanics, the axial direction is 0.1 mm/ min, and the strain monitoring is kept synchronized during the loading process. We adopt the strain resistance method for strain monitoring. Through the bridge of the strain resistance meter, the measured resistance value is converted into the strain value. We also select the resistance strain gages with a size of 18 mm × 7 mm and an insulation resistance of 5,000 mΩ. On the 4 directions in the middle of each sample, a total of 8 lateral and vertical resistance strain gages are symmetrically attached. The gages should be firmly attached to avoid any cracks, as shown in Figure 4.2.

 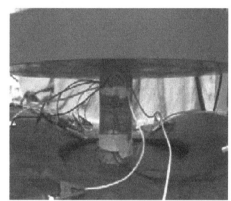

Figure 4.2   Single-axis compression test scene

2) Experimental results and analysis

To a certain extent, the test results of rock strength will be affected by many factors, such as the sampling location, the anisotropy of the rock itself, the diversity of the mineral composition, the loading rate, and some artificial factors, etc. The results of uniaxial compression tests on andesite and gneiss samples are shown in Table 4.1, and the stress-strain curve of the entire uniaxial compression process is shown in Figure 4.3.

Table 4.1   Uniaxial compression test results

Lithology	Dry compressive strength /MPa	Wet compressive strength /MPa	of Poisson Ratio	Elastic Modulus /GPa
Andesite	138	104	0.26	42.5
	120	101	0.29	41.2
	112	95	0.32	38.3

Continued

Lithology	Dry compressive strength /MPa	Wet compressive strength /MPa	of Poisson Ratio	Elastic Modulus /GPa
Gneiss	143.8	127	0.20	56.1
	141.4	118	0.21	56.3
	142.2	121	0.20	56.1

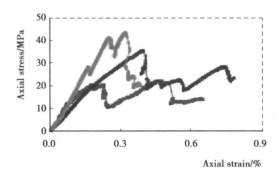

Figure 4.3 Axial stress-strain curve of uniaxial compression

4.2.2 Tensile strength test

The tensile strength of a rock refers to the maximum stress value that the rock can withstand under tensile conditions. Through the laboratory tests, the tensile strength of the rock is obtained, and compared with the rock's compressive strength, the strength characteristics of the rock can be obtained.

1) Experimental process

In this test, the Brazilian splitting method, one of the indirect methods, is used to determine the tensile strength of the rock. The splitting method is to apply an external load to the sample to produce tensile stress inside it, and then the amount of stress can be calculated based on the elastic theory. As shown in Figure 4.4, the upper left corner is the rock samples, the right side is the scene of test process, and the lower left corner is the results of the Brazilian splitting method. The test equipment is the same as used in the uniaxial compression test, that is, the microcomputer-controlled pressure tester and the static resistance meter.

2) Experimental result analysis

The specific results of the Brazilian splitting test are shown in Table 4.2. The results indicate that:

The dry tensile strength of andesite in the tunnel site is between 13 and 16.99 MPa, while the wet tensile strength is between 9 and 11 MPa. The dry tensile strength of gneiss is between 15 and 16.5 MPa, while the wet tensile strength is between 10.3 and 11.5 MPa. The changing rule is similar to that of the compressive strength.

After the sample is stretched and splited, the fracture site is obvious and uneven.

Figure 4.4 Brazil split test scene

Table 4.2 Rock tensile strength test table

Lithology	Dry tensile strength/MPa	Wet tensile strength/MPa
Andesite	15.7	10.1
	15.4	9.4
	14.3	10.7
Gneiss	16.5	11.2
	16.1	10.6
	15.8	10.3

4.2.3 Normal triaxial compression test

Since most rocks in nature are in a three-way compressive stress state, which is a typical stress state, the strength deformation characteristics of the rock under the three-way compressive stress can, in a sense, truly reflect the mechanical properties of the rock. Therefore, studying the strength and deformation characteristics of rocks under three-way stress is of great significance to the geotechnical engineering (Qian Jiahuan et al., 1996).

The triaxial mechanical test uses the "large-scale high-pressure rock permeability tester", developed independently by State Key Laboratory of Geohazard Prevention and Geoenvironment Protection, as shown in Figure 4.5. Controlled by a computer, this tester has 3 operation modes: the full-automatic mode, the semi-automatic mode and the manual loading mode. Through the mutual regulation of electricity, gas and liquid, it can accurately control the confining pressure, the axial pressure and the axial displacement. It can be used to do triaxial mechanics tests on rock pieces of various sizes. The maximum confining stress of the tester is 30 MPa and the maximum axial stress is 4,000 kN.

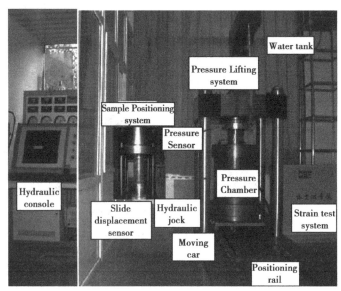

Figure 4.5 Large-scale high-pressure rock permeability tester

Researchers of this experiment apply the triaxial compression test under the condition of equal lateral pressure, that is, a normal triaxial compression test, to obtain those parameters such as the lateral stress, the axial stress, the axial displacement, and the time. By processing these data, we can get such mechanical parameters as the yield stress, the peak stress, the axial strain and the stress difference, and even the related relation curves.

1) Experimental process

According to the regional geological data of the studied area, the maximum confining pressures for both kinds of rock samples is set as 15 MPa, and the side pressure stages are evenly divided, according to the number of samples, that is, the confining pressures are set to 5 MPa, 10 MPa, and 15 respectively.

A pressure of 2.5 MPa/min is applied to have a constant pressure loading control. After reaching the preset confining pressure value, the confining pressure is kept unchanged, and the axial displacement control with a loading rate of 0.1 mm/min is used instead. When the axial displacement change rate is 0.3 mm/s, the test ends.

2) Experimental results and analysis

By the triaxial test method, the peak strengths of the rock under different confining pressure conditions can be obtained, and further the axial stress σ_1 and the lateral stress σ_3 during failure can also be obtained. When σ_3 is 5 MPa, 10 MPa, and 15 MPa respectively, the conventional triaxial strength parameters of andesite and gneiss in the tunnel site are shown in Table 4.3. Based on the conventional triaxial strength parameters of andesite, the Mohr stress circle can be drawn on the coordinate plots for the shear force τ and the normal stress σ, and then based on the least square method, the relation curve between σ_1 and σ_3 can be drawn, as shown in Figure 4.6. Finally, by using the relation curve and equation (4.1) of Mohr stress circle ($\sigma_1-\sigma_3$), the cohesive force, the internal friction angles, the compressive strength and the tensile strength of andesite and gneiss can be calculated.

Table 4.3 Conventional triaxial lateral stress and axial stress and their ratios

Number	σ_3/MPa	σ_1/MPa	$(\sigma_1-\sigma_3)$/MPa	σ_1/σ_3
E5-1	5	115.1	110.1	23.02
E5-2	10	197.5	187.5	19.75
E5-3	15	268.4	253.5	17.89
S5-1	5	125.1	120.1	25.02
S5-2	10	199.7	189.7	18.97
S5-3	15	257.1	242.1	16.14

$$\begin{cases} R_t = \dfrac{2c \cos \varphi}{1 + \sin \varphi} \\ R_c = \dfrac{2c \cos \varphi}{1 - \sin \varphi} \\ \sigma_1 = R_c + \dfrac{1 + c \cos \varphi}{1 - \sin \varphi} \sigma_3 \\ c = \dfrac{R_c(1 - \sin \varphi)}{2 \cos \varphi} \end{cases} \quad (4.1)$$

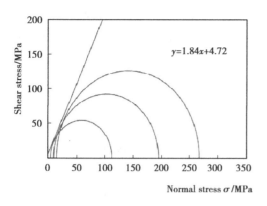 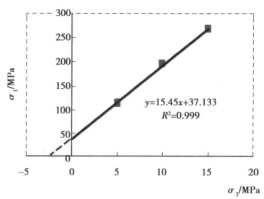

Figure 4.6　Mohre stress circle and (σ_1-σ_3) relation curve

Where c is cohesive force of rock (MPa), φ is internal friction angle of rock (°), R_c is intercept of the best relation curve on the ordinate, which is equivalent to the compressive strength of rock (MPa), and R_t is tensile strength of rock (MPa).

From the results of the uniaxial compression test, we find that the average uniaxial compressive strength of andesite is 121 MPa and the average tensile strength is 10.4 MPa. Calculated from the triaxial tests, the cohesive force of andesite is 20.2 MPa, the internal friction angle is 44°, and the compressive tensile strength is 6.40 MPa. As to gneiss, the average uniaxial compressive strength is 143.8 MPa, and the average tensile strength is 12.3 MPa. Calculated from the triaxial tests, the cohesive force of andesite is 23.1 MPa, the internal friction angle is 48°, and the compressive tensile strength is 7.80 MPa. Comparing the results of the two tests, we find the compressive strengths of the two rocks are not much different, so they can check against each other.

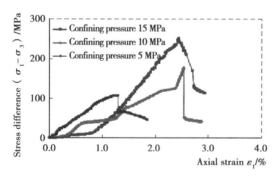

Figure 4.7　Stress-strain curves under of different surrounding rocks

Figure 4.7 shows the stress-strain curves of different surrounding rocks, from which we find that the stress-strain curve of andesite rock can be divided into three stages, or three sections, each of which can be explained separately by conventional theories (Hong Kairong, 2015).

①The compaction stage: At the initial stage of loading, the sample is under smaller pressure, so the deformation is mainly caused by the discontinuous closures, such as the internal joints, the cracks etc., and the compression of the filler. This is called the compaction stage. At this stage, some large deformations occur under relatively small stress. At the same time, when the direct stress is low, the friction between the cracks is so small that it may lead to slippage or greater axial deformation in the rock sample. The unevenness of the ends of piece and the unevenness of the indenter ball seat will also cause some non-linear changes in the initial stage, so the stress-strain relation curve is concave, and the deformation modulus is small.

②The elasticity stage: As the rock sample is fully compacted, the structure in the sample begins to be compressed. In this process, the rock sample is undergoing elastic structural deformation, thus it is called the elasticity stage. The stress-strain curve is linear and the deformation modulus is relatively large.

③The plasticity stage: As the pressure continues to increase, after the elastic strength limit of the rock is exceeded, new cracks will be generated and extend inside the rock sample, and large deformation will occur with even small increase in compressive stress, so the rock sample enters the plasticity phase. Owing to the strong integrity of the rock sample, the plasticity deformation is not so large and the ductility is not so obvious.

Based on the previous three tests, and consulting the "Initial Investigation Report of the Erlang Mountain Tunnel on the Yakang Expressway" and the "Report on the Survey and Design of the Sangzhuling tunnel of the Linla Railway", we obtain the physical and mechanical parameters of the andesite and the gneiss as shown in Table 4.4.

Table 4.4 Rock mechanical parameters

Lithology	Elastic Modulus /GPa	Poisson's ratio	Cohesion /MPa	Internal friction angle/(°)	Compressive strength/MPa
Andesite	40	0.3	20.2	44	100
Gneiss	56	0.20	23.1	48	143.8

In addition, a density test is also performed on the rock samples. The density of the andesite is 2,650 kg/m^3 and the density of gneiss is 2,680 kg/m^3. Because the experimental process is simple, we will not repeat them in this part.

4.3 Summary

This chapter mainly introduces the process of laboratory test on the rock samples collected in Erlangshan Tunnel and Sangzhuling Tunnel to obtain the original rock mechanical parameters for the requirement of simulation calculations. By performing the uniaxial compression deformation test, the normal triaxial compression test, the tensile strength test on rock samples, and calculating the rock deformation parameters, we finally obtain the rock mechanical parameters required for calculating and analyzing samples in this book: as to the andesite, the compressive strength is 100 MPa, the elastic modulus 40 GPa, the cohesive force 20.2 MPa, the Poisson's ratio 0.3, the internal friction angle 44°, the density 2,650 kg/m^3; while for gneiss, the compressive strength is 143.8 MPa, the elastic modulus 56 GPa, the cohessive force 23.1 MPa, the Poisson's ratio 0.20, the internal friction angle 44°, and the density 2,680 kg/m^3.

Chapter 5 Optimization of Parameters for Pre-release De-stress Controlled Blasting Based on Numerical Simultations

5.1 Response Platforms of Numerical Simulation Analysis

In engineering blasting, the explosives release high energy in an extremely short time, and instantly act on the surrounding medium, destroying its internal structure. This sudden change of load puts the entire structure in a severe dynamic motion, which brings very complicated force and response to the structure itself. To study this process, we need the help of the structural dynamics theories. In this high in-situ stress relief blasting, ANSYS/ LS-DYNA software is selected to simulate and analyze the blasting effect of explosives on surrounding rocks at an instant. The main reason is that ANSYS/ LS-DYNA, as the world's most famous all-purpose explicit dynamic analysis program, can simulate various complex situations in the real world, and is particularly suitable for solving various non-linear dynamic problems, such as high-speed collisions, explosions and metal forming in 2-dimensional or 3-dimensional structures.

The impact range of the high-speed explosion is mainly concentrated in the crushing zone, the crack zone, etc. So it is no advantage to use the ANSYS/ LS-DYNA software to analyze the external mechanical action. Since the stress in the surrounding rocks still change within a certain time, after the blasting pressure is released, it is proposed to use FLAC 3D numerical analysis software to accurately analyze this trend. In summary, the ANSYS/ LS-DYNA software is used to simulate the extreamely close zone, while outside the extreamely close zone, the blasting stress calculated by ANSYS/ LS-DYNA is derived as load to put in the FLAC 3D, so as to establish a response platform for the effect analysis of de-stress control blasting under high geostress conditions.

5.1.1 ANSYS/LS-DYNA numerical platform

The ANSYS/LS-DYNA numerical platform has the following features and functions:
①Features of ANSYS/LS-DYNA software analysis: As a three-dimensional high-speed and

highly nonlinear analysis program with explicit finite element, ANSYS/LS-DYNA971 uses some advanced and sufficient numerical processing technologies and thus has some distinctive technical features, such as explicit algorithms and time-step control, rich material models, simple and applicable element types, and a multiple ways of contacting and coupling, all of which having been successfully applied in the research of the vehicle collisions, the airbag and seat belt systems, the metal forming of thin plates in contact with punches and abrasive tools, the effects of underwater explosions on structures, and the calculation of the penetration of high-speed projectiles on target plates.

②Features of ANSYS/LS-DYNA algorithm: ANSYS/LS-DYNA has three basic algorithms, namely the Lagrange algorithm, the Euler algorithm and the ALE algorithm. The features of Lagrange algorithm is that the mesh is attached to the material, and the element mesh deforms with the flow of the material; but when the structural deformation is too large, it may cause serious distortion of the finite element mesh, causing numerical calculation difficulties, and even the termination of calculation program. In contrast, the ALE algorithm can effectively overcome the severe distortion of the units and achieve the dynamic analysis of fluid-solid coupling. The ALE algorithm first performs one or several Lagrange time-step calculations. At this time, the element mesh is deformed with the flow of the material, and then the ALE time-step calculation is performed, and its starting time, end time and frequency can be selected by itself. In the Euler algorithm, the material is flowing in a fixed grid. The Euler algorithm has obvious advantages for model fluids, including air and water, and can make the calculation results closer to the real situation. The LS-DYNA can easily couple the Euler grids with the ALE grids to deal with the interaction of fluids and structures under various complex load conditions.

5.1.2 FLAC 3D numerical analysis platform

FLAC 3D, a professional numerical analysis software for geotechnical engineering developed by Itasca International Group, is widely used in slopes, foundation pits, tunnels, underground caverns, mining, energy and nuclear waste storage. FLAC 3D can be used to calculate the deformation, stress, stability of rocks and soil bodies under various external loads, and the large deformation after post-peak and its characteristics. FLAC 3D needs to carry out the CONFIG dynamic command before dynamic analysis. Since the FLAC 3D can simulate the complete nonlinear response of rock and soil under external (such as earthquake) or internal (such as wind, explosion, subway vibration) loads, it can be applied to the calculations in rock mechanics and other disciplines.

Taking a complete non-linear analysis method, and based on the explicit difference analysis, FLAC 3D uses the concentrated mass of grid nodes obtained from the real density of surrounding rocks to derive all motion equations. Thus FLAC 3D can follow any specified non-linear constitutive model. By using FLAC 3D, waves of different frequencies can interfere and mix naturally, based on some non-linear material rules; thus permanent deformation can be automatically calculated. A reasonable plastic equation can also be used to make the plastic strain increase related to the stress.

5.2 Calculation Process of De-stress Controlled Blasting

5.2.1 ANSYS/LS-DYNA numerical simulation blasting

The numerical simulations using different finite element software are generally similar in the calculation process, which is usually divided into three steps: preprocessing, solving and the postprocessing. The simulation process of ANSYS/ LS-DYNA also follows the above three steps, and its workflow and file system are shown in Figure 5.1.

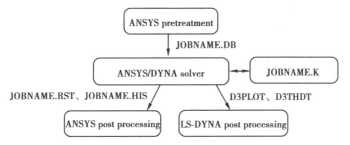

Figure 5.1 ANSYS/ LS-DYNA calculation process and file system

(1) Difficulties of ANSYS/ LS-DYNA numerical simulation

Before solving, it is necessary to modify the keyword file, which is unique to LS-DYNA analysis, and it is also the most difficult point of analysis. The K files are the information files that are output through preprocessing after the model is created.

It contains all the information needed to the problem, such as nodes, element information, materials and equations of state, contact, conditions of initial boundary value, and load information, which are all expressed in the format of LS-DYNA keyword commands beginning with " * ", and the solver will follow the commands in the K file when solving.

In most cases, the K file, automatically generated by ANSYS pre-process, requires some artificial modification. This is because a large part of the LS-DYNA function of ANSYS pre-process cannot be directly implemented. For example, in term of material, ANSYS cannot provide the types and the constitutive parameters of some material unique to LS-DYNA; instead it needs to be manually added from the K file. In the subsequent numerical simulation of the de-stress of explosives, it is through the modification of the K file that we can restart the analysis and deal with some difficulties in ANSYS. After the modification of K file, the solver can read the information for calculation.

(2) Basic principles of modeling

There are two basic principles of modeling: the principle of ensuring accuracy and the principle of controlling scale. The two are usually contradictory. If the accuracy is high, the model scale generally increases; while if the model scale decreases, the accuracy decreases accordingly. Therefore, we need to weigh and make a comprehensive decision according to the specific objects, requirements and conditions when modeling.

(3) Establishment of model assumptions

When establishing a numerical model for de-stress controlled blasting, the main task is to obtain the shelling force on the blasthole wall. Therefore, the blasting effect is the main factor in this simulation analysis. In order to decrease the difficulty in establishing and analyzing the numerical model, some minor factors need to be ignored, and thus the following assumptions are made about the model:

①The rock is assumed to be an isotropic continuous homogeneous medium, and the expansion of the detonation products is an adiabatic process;

②Since the effect of gravity is small relative to the detonation pressure, the effect of gravity is not considered.

(4) Pre-processing

The modeling and loading process of ANSYS/ LS-DYNA is performed in the environment of ANSYS pre-processing, and the basic operation steps are as following: first add several options of LS-DYNA, including the definition of element type, the definition of material type, and the element properties; then establish geometric finite element models, divide finite element meshes, to create PART sets, define constraints, apply boundary conditions and loads, and finally set the model output options, including calculation time, steps of calculation time, solutions, and types of output files.

①Model unit system. Since the explosion is a transient process, from the initiation of explosives to the completion of explosive explosions, they are all in a subtle level (μs). In order to reduce the calculation errors and improve the calculation accuracy, it is decided to use the cm-g-μs unit system. So all parameters need to be converted to this unit system before using.

②Model unit selection. The choice of unit during modeling is directly related to the method of modeling. This modeling takes a bottom-up approach: first generate the points, lines, and faces, and then extrude the faces into some elements. Therefore, in selecting the surface element and the body element, we choose the surface element SHELL 163, and the body element SOLID 164, as shown in Figure 5.2.

Chapter 5 Optimization of Parameters for Pre-release De-stress Controlled Blasting Based on Numerical Simultations

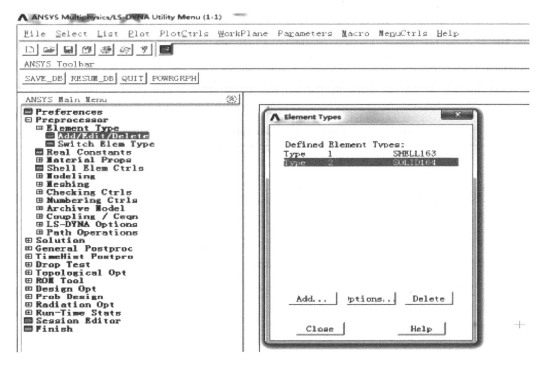

Figure 5.2 Model unit selection

③Selection of material models and state equations.

Ⅰ. Explosives model and state equation

In ANSYS/ LS-DYNA numerical simulation, the material model of explosives takes only the high-energy explosive model HIGH_EXPLOSIVE_BURN, and has the following relationship:

$$f = \max(f_1, f_2) \qquad (5.1)$$

$$f_1 = \begin{cases} \dfrac{2t - t_e D}{3V_e} \\ \dfrac{A_{max}}{A_{max}} & t > t_e \\ 0 & t \leqslant t_e \end{cases} \qquad (5.2)$$

$$f_2 = \frac{1 - V}{1 - V_{CJ}} \qquad (5.3)$$

Where t_e is the shortest time required for the explosive stress wave to reach the centroid of the current unit, f is the combustion coefficient, if $f>1$, $f = 1$.

As to the state equation, we choose JWL to simulate the relationship between pressure and specific volume during the explosion process (the only state equation that matches the explosive model in ANSYS). The relationship is as follows:

$$P = A\left(1 - \frac{\omega}{R_1 V}\right) e^{-R_1 V} + B\left(1 - \frac{\omega}{R_2 V}\right) e^{-R_2 V} + \frac{\omega E_0}{V} \qquad (5.4)$$

Where V is the relative volume of the detonation product, which is equal to the ratio of the volume of the explosive products to the initial volume; E_0 is the initial specific internal energy, 105 MPa; ω is the Green Eisen parameter; A, B are constant; R_1, R_2 are dimensionless parameter.

The specific parameters of the main explosives (Zhu Liang et al., 2011) are shown in Table 5.1.

Table 5.1 Explosive material parameters and state equations

$\rho/(\mathrm{kg}\cdot\mathrm{m}^{-3})$	$D/(\mathrm{m}\cdot\mathrm{s}^{-1})$	P_q/GPa	A/GPa	B/GPa	R_1	R_2	ω	E_0/GPa
1,630	6,930	27.0	371	74.3	4.15	0.95	0.3	1.0

II. Air material model and state equation

The air material model takes the NULL model (Figure 5.3), and the equation of state is described by using the LINEAR_POLYNOMIAL model. The internal energy of this equation is linearly distributed, and the pressure is given by:

Figure 5.3 Air material model selection

$$P = C_0 + C_1\mu + C_2\mu^2 + C_3\mu^3 + (C_4 + C_5\mu + C_6\mu^2)E \quad (5.5)$$

Where P is detonation pressure, E is internal energy per unit volume, V is relative volume, and C_i is linear polynomial state equation constant ($i = 1, 2, 3, 4, 5, 6$).

When the linear polynomial equation of state is used in the air model:

$$C_0 = C_1 = C_2 = C_3 = C_6 = 0, \ C_4 = C_5 = 0.4$$

The specific parameters of air material are shown in Table 5.2.

Table 5.2 Air material model and state equation parameters

$\rho/(\text{kg}\cdot\text{m}^{-3})$	C_0	C_1	C_2	C_3	C_4	C_5	C_6	E_0/Pa	V_0
1.29	0	0	0	0	0.4	0.4	0	2.5	1

Ⅲ. Rock models and material parameters

In practice, the deformation of surrounding rocks in high geostress situations can be elastic deformation, plastic deformation, elastoplastic deformation and other complex deformation. The nonlinear plastic follow-up strengthening material model of DYNA is applicable to all the follow-up strengthening materials (Figure 5.4). Therefore, PLACTIC_KINEMATIC is selected as the material model of the surrounding rocks. The follow-up reinforcement model is built on the Cowper-Symonds relationship, and its expression is as follows:

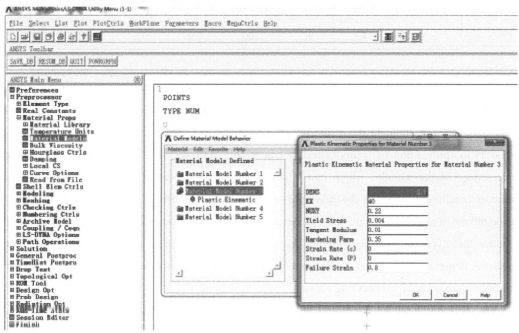

Figure 5.4 Rock material model selection

$$\sigma_y = \left[1 + \left(\frac{\dot{\varepsilon}}{C}\right)^{\frac{1}{p}}\right](\sigma_0 + \beta E_p \varepsilon_{\text{eff}}^p) \qquad (5.6)$$

Where σ_0 is initial yield strength, C, P are material-related constants, and β is adjustable parameters. When $\beta = 0$, it means the plastic follow-up strengthening model, and when $\beta = 1$, it is the isotropic strengthening model.

Here E_p is a plastic strengthening model, and its expression is:

$$E_p = \frac{E_t E}{E - E_t} \qquad (5.7)$$

$\varepsilon_{\text{eff}}^p$, the equivalent plastic strain is calculated as:

$$\varepsilon_{\text{eff}}^p = \int_0^t \left(\frac{2}{3}\dot{\varepsilon}_{ij}^p \dot{\varepsilon}_{ij}^p\right)^{\frac{1}{2}} \mathrm{d}t \tag{5.8}$$

Where $\dot{\varepsilon}_{ij}^p$ is the plastic strain rate of the material, and its value is equal to the total strain rate minus the elastic strain rate.

The main parameters in the material model include: density (ρ), elastic modulus (E), Poisson's ratio (ν), yield strength (σ_Y), tangent modulus (E_t), strengthening parameter (β), failure strain (F_s), etc. According to the experimental results in Chapter 4, and based on the detailed survey report of Erlangshan Tunnel on Yakang Expressway, the specific parameters of andesite are shown in Table 5.3.

Table 5.3 **Rock material model parameters**

$\rho/(\text{kg} \cdot \text{m}^{-3})$	E/GPa	ν	σ_Y/MPa	E_t/GPa	β	F_s
2,650	40	0.3	100	4.0	0.5	0.8

④ Boundary conditions of models. First of all, for the boundary conditions of models, normal displacement constraints are imposed on the three outer surfaces of the model X, Y and Z (Figure 5.5). Secondly, to simulate the explosion of explosives in infinite surrounding rock mass, it is necessary to apply a non-reflecting boundary condition (NON-REFLECTING BOUNDARY) on the outer surface of the models, which can effectively reduce the size of model. The non-reflection boundary condition, based on the principle of virtual work, converts the distributed damping on the boundary into an equivalent node force and adds it to the boundary, that is, to list all the elements that make up the non-reflection boundary, and to add the viscosity stress and shear stress on all the non-reflection elements. The non-reflection boundary conditions can be used to simulate the infinite situation of surrounding rocks.

For the force boundary conditions of the model, since the main task of the DYNA platform is to simulate and calculate the shelling force of the explosive on the blasthole wall, so that there is no need to add other external loads, that is, no need to add force boundary conditions.

⑤ Setting of analysis option to output and modify K file (Figure 5.6). Before generating the K file, we need to set the analysis options, including the calculation energy type, the solution time and step length, the output type and the time interval, etc.

After the K file is generated, keywords in the K file need to be added, deleted, or modified. The keywords include:

∗SECTION_SOLID_ALE, ∗ALE_MULTI-MATERIAL_GROUP, ∗CONSTRAINED_ LAGRANGE_IN_SOLID, ∗MAT_HIGH_EXPLOSIVE_BURN, ∗CONTROL_ALE, ∗INITIAL_ DETONATION, ∗EOS_LINEAR_POLYNOMIAL, ∗CONTROL_TIMESTEP, ∗SET_PART_

Chapter 5 Optimization of Parameters for Pre-release De-stress Controlled Blasting Based on Numerical Simultations

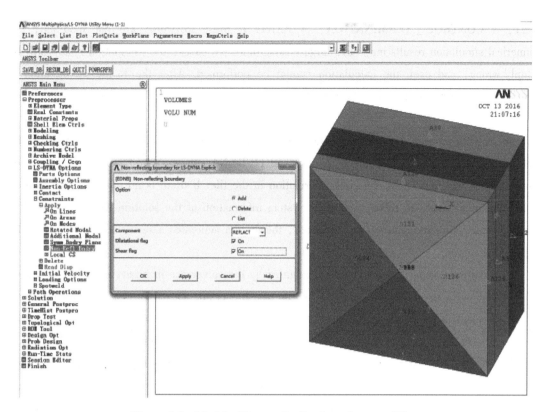

Figure 5.5 Model with no reflection boundary condition

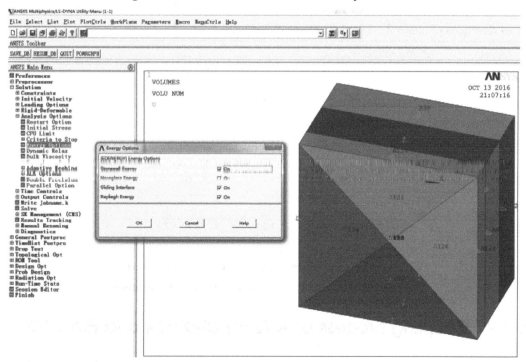

Figure 5.6 Setting K file analysis options

LIST, *CONTROL_TERMINATION, *PART. If these keywords are not modified, the numerical simulation results may have large errors. In severe cases, the calculation results will be simply wrong, and even the calculation cannot be performed. After the K file modification is completed, import the K file into LS-DYNA Solver for calculation and save the calculation results.

⑥ Solving process. For the convenience of control, LS-DYNA's solution process is generally not performed in ANSYS, but solved by a special LS-DYNA solver. After LS-DYNA Solver reads the K file, it will first check the error information in the file. If there is no mistake, the solution will automatically start to solve, and some system information of the solution will be provided to users in time. After the solution is completed, the system generates a series of result files in the solution directory.

⑦ Post-processing. The result files generated by the solver are in binary machine language, which needs to be displayed and analyzed by the post-processing software. As a very important part of a finite element program, the post-processing software must be able to fully support the solution results. LS-DYNA's post-processing software generally uses a special post-processing software LS-PREPOST. After reading the result files, the post-processing software can intuitively output information, such as the stress-strain cloud diagram, the time history curve, and the result animation. In the LS-Pre Post post-processing analysis, the calculation results of Solver are opened by the LS-Pre Post post-processor to extract the speed, acceleration, and other information in the three directions of nodes X, Y and Z (Figure 5.7–Figure 5.9).

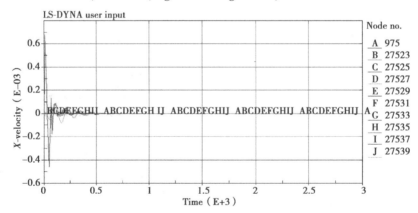

Figure 5.7　Ten particles velocity extraction in the X direction
of the first group of blast holes in the orthogonal experiment

5.2.2　Coupling process of ANSYS/LS-DYNA and FLAC 3D

The results obtained by ANSYS/LS-DYNA are used as dynamic loads, which are applied to the FLAC 3D software conditions to make FLAC 3D nonlinear dynamic calculation. When using

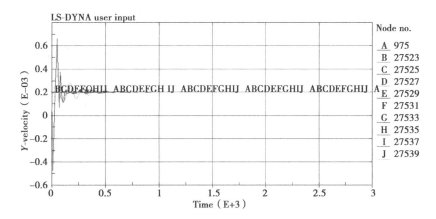

Figure 5.8 Ten particles velocity extraction in the Y direction
of the first group of blast holes in the orthogonal experiment

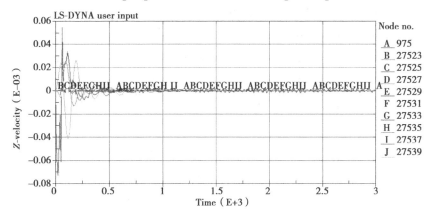

Figure 5.9 Ten particles velocity extraction in the Z direction
of the first group of blast holes in the orthogonal experiment

FLAC 3D to make nonlinear calculations, there are three issues to be noticed:

①Application of dynamic loads. FLAC 3D can apply dynamic loads at the model boundaries or internal nodes to simulate the response of materials to external or internal dynamic forces. The dynamic load inputs allowed by the program can be: the acceleration time history, the speed time history, the stress (pressure) time history and time schedule of concentrated force. The application of dynamic loads requires the APPLY command and the use of the table command to input non-equidistant dynamic loads, shown in Figure 5.10. In FLAC 3D dynamic calculation, because ANSYS extracts more node information and uses a larger amount of command inputs, it is necessary to use the method of editing text files to read the tables.

② Dynamic boundary conditions. As to dynamic problems, the selection of boundary conditions around the model is very important, because there will be wave reflections on the boundary, which may affect the results of dynamic analysis. The larger the scope of the analysis model is, the better the analysis results will be; but a larger model will cause a huge

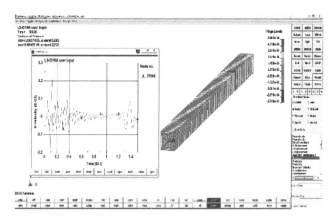

Figure 5.10 Maximum principal stress of the tunnel after de-stress
blasting in the first group of orthogonal experiments

computational burden. The FLAC 3D provides two boundary conditions: the static (viscous) boundary and the free field boundary to reduce the wave reflection on the model boundary.

For the case where the dynamic load originates from the inside of the model, the dynamic load can be directly applied to the node. In this case, using the static boundary can effectively reduce the reflection on the artificial boundary, and it is unnecessary to apply the free field boundary mentioned below.

③Mechanical damping. Damping is mainly caused by the internal friction of the material or the possible sliding of the contact surfaces. The FLAC 3D uses the method of solving dynamic equations to solve two types of mechanical problems: the quasi-static problems and the dynamic problems. Damping is used in both types of problems, but the quasi-static problems require more damping to allow dynamic equations to converge and balance as fast as possible. For damping in dynamic problems, it is necessary to reproduce the damping magnitude of natural system under dynamic load in numerical simulation. At present, the FLAC 3D dynamic calculation provides three types of damping for users to choose: Rayleigh damping, local damping and hysteretic damping. This simulation uses Rayleigh damping.

The mass component in Rayleigh damping is equivalent to the damper connecting each node to the ground, and the stiffness component is equivalent to the damper between the connecting elements. Although the two dampers are frequency-dependent, by selecting appropriate coefficients, a frequency-independent response can be obtained approximately within a limited range of frequency. Rayleigh damping can approximately reflect the frequency independence of rock and soil.

The two parameters in Rayleigh damping are the minimum critical damping ratio ξ_{min} and the minimum center frequency ω_{min}. For geotechnical materials, the range of critical damping ratio is generally 2%-5%.

5.2.3 Stress release simulation of de-stress blasting in tunnel

1) Determination of levels of surrounding rocks in tunnel

The classification of the surrounding rocks in tunnel is based on the calculation formula of the basic quality index of rock mass in the "Code for Design of Highway Tunnels" (JTG D70-2004):

$$BQ = 90 + 3R_c + 250K_v \tag{5.9}$$

When $R_c \geqslant 90K_v + 30$, take the smaller value. When $K_v \geqslant 0.04R_c + 0.4$, take the smaller value. Then substitute the above formula to calculate the revised value of the basic quality index of BQ rock mass:

$$[BQ] = BQ - 100 \times (K_1 + K_2 + K_3) \tag{5.10}$$

Check the specifications and integrate the previous analysis, and the surrounding rocks' classifications in the tunnel can be calculated as in Table 5.4.

Table 5.4 calculation table of classification of tunnel surrounding rocks

Geotechnical category	Status	R_c/MPa	K_v	BQ	Correction factor		[BQ]	Surrounding rock level
					K_1	K_2		
Limestone	Medium to thick layered	68.9	0.55	434.2	0.2	0.5	364.2	III
Quartz sandstone, quartz siltstone	Medium to thick layered	79.5*	0.55	466	0.2	0.2	426	III
Limestone, quartz sandstone and mudstone	Medium to thick layered	79.5*	0.55	466	0.2	0.5	396	III
Granite	Lump	81.2	0.62	488.6	0.2	0.1	458.6	II
	Fissure block	81.2	0.55	471.1	0.1	0.3	431.1	III
Andesite	Fissure block	78	0.55	461.5	0.1	0.3	421.5	III
Fine grain diorite	Lump	85.8*	0.62	502.4	0.2	0.1	472.4	II
Granite mixed rock	Fissure block	75*	0.52	445	0.2	0.3	395	III
Limestone shale, mudstone	Thin to medium thick layered	47**	0.48	343	0.3	0.2	293	IV
Sandstone with shale, mudstone	Thin to medium thick layered	34**	0.48	312	0.3	0.2	262	IV

Continued

Geotechnical category	Status	R_c/MPa	K_v	BQ	Correction factor		[BQ]	Surrounding rock level
					K_1	K_2		
Thin to medium thick mudstone, shale	Thin to medium thick layered	14.3	0.48	253			253	IV
Medium to thick layered mudstone	Medium to thick layered	16.7	0.48	260			260	IV
Fault fracture zone	Tectonic rock							V
Overlay	Bulk structure							V

2) The characteristics of the initial stress data of the tunnel in-situ stress field simulation

The Erlangshan Tunnel is located at the WS end of the NE trending structural belt of Longmen Mountain, and the tunnel site is located at the NE side of the junction of "Y" type structural fork, and the overall tectonic framework is controlled by the NE trending of Longmen Mountain structural belt.

In the Erlangshan Tunnel, the CZK2, CZK3, and CZK4 sites are tested with the hydraulic fracturing method. Among them, CZK2 and CZK3 are located at the top of the ridge and their depth is greater than 600 m, while CZK4 is about 460 m in depth, mainly located in the residual stress zone and the geostress relaxation bands; so we don't analyze it here. By analyzing the test data of CZK2 and CZK3, we can fit the changing line of the geostress with the increase of depth in the tunnel site. The details are shown in Figure 5.11.

According to the fitting formula of maximum principal stress of CZK3, the maximum principal stress at a buried depth of 751 m is 30.25 MPa, which, though, is somewhat different from the simulation result, but since the difference is less than 15%, it can still be used as the basis for analyzing the geostress in large sections of buried depth. According to the CZK2 simulation formula, the maximum principal stress is 23.9 MPa at a depth of 750 m, and then at a maximum depth of 1,590 m, the maximum principal stress is 43.05 MPa, and the minimum principal stress is 27.55. Using the CZK3 test to simulate the stress field in the tunnel site, after calculations, we find that when the maximum buried depth of the tunnel site is about 1,590 m, the maximum

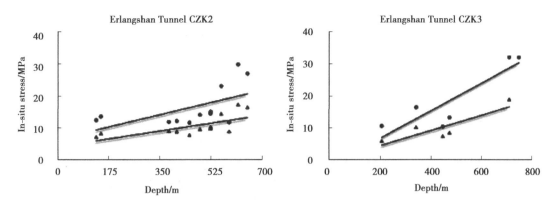

Figure 5.11 Erlangshan Tunnel CZK2 and CZK3 geostress regression diagram

principal stress is 65.8 MPa, and the minimum principal stress is 36.68 MPa.

Meanwhile, in the construction of Erlangshan Tunnel, applying the stress relief method in the tunnel, we find that when the tunnel is buried at a depth of 750 m, the maximum principal geostress is 35.3 MPa and the direction is NW85°; the intermediate principal stress is 15.3 MPa, and the minimum principal stress is 8.1 MPa.

Furthermore, based on the field test of rockbursts and related large deformations in the Nibashan of the Yaxi Expressway, by analyzing a large number of geostress measurement data at home and abroad, the research group obtain the relational formula of geostress after sampling a total of 276 pieces of field geostress measurement data around the world. The relational formula can be applicable in a depth range of 0–5,000 m. The specific regression fittings formulas are as follows:

$$\sigma_v = 1.88 + 0.244H \quad (5.11)$$

$$\sigma_{hmax} = 13.763 + 0.0211H \quad (5.12)$$

$$\sigma_{hmin} = 6.464 + 0.0146H \quad (5.13)$$

Where σ_v is vertical geostress, σ_{hmax} is maximum horizontal geostress, and σ_{hmin} is minimum horizontal geostress.

It can be predicted that the maximum principal stress is 47.30 MPa and the minimum principal stress is 29.67 MPa. To sum up, the maximum principal stress in the tunnel site can be selected as the average of the three predicted results, i.e. the maximum principal stress is 51.6 MPa, and the minimum principal stress is 31.3 MPa. Combined with the forecast results of various methods, it can be inferred that the main geostress values in the tunnel are shown in Table 5.5:

Table 5.5 Tunnel geostress forecast table

Method of forecast	item	1,590 m	1,500 m	1,200 m	1,000 m	800 m	600 m
G318 CZK2	σ_1	43.05	41.00	34.16	29.60	25.04	20.48
	σ_3	27.56	26.25	21.86	18.94	16.02	13.10
G318 CZK3	σ_1	66.45	62.58	49.64	41.02	32.40	23.78
	σ_3	37.21	35.07	27.96	23.22	18.48	13.74
Nibashan Tunnel	σ_1	47.31	45.41	39.08	34.86	30.64	26.42
	σ_3	29.68	28.36	23.98	21.06	18.14	15.22
Average value	σ_1	51.6	49	40.8	35.1	29.2	23.6
	σ_3	31.3	29.6	24.5	21.06	17.5	14

The existing geostress analysis results are mainly based on the tests carried out with the deep hole hydraulic fracturing method. While the actual excavation verification shows that the test value of the hydraulic fracturing method is about 5 – 10 MPa higher than that of the in-hole stress relief method; and the intermediate principal stress, estimated in the previous period, is vertical, which is different from the actual geostress field. Therefore, it is necessary to enforce the on-site geostress measurements during the on-site excavation so as to prove the preliminary estimation results.

The maximum principal stress direction is NW 65° – 85°. Combined with the tunnel axis direction, the maximum stress in the tunnel's normal direction can be determined to be 43.7 MPa. Its ratio to the lithologically saturated uniaxial compressive strength of the tunnel site is less than 4, which indicates that it is an extremely high geostress tunnel site, which means in the excavation process, the rock bursts might occur in hard rocks, and large deformation might occur in weak rocks.

3) FLAC 3D computing tunnel model

(1) Model size

In this simulation, the calculation area is selected at the K6+054 to K76+280 andesite zone in New Erlangshan Highway Tunnel. The calculation area is 50 m long, 20 m wide and 50 m high. The circular tunnel is located in the middle of the model and has a radius of 5 m. After the tunnel is excavated for a length of 5 m, the FLAC 3D mesh is as shown in Figure 5.12. The calculation model has 160,000 units and 164,041 nodes.

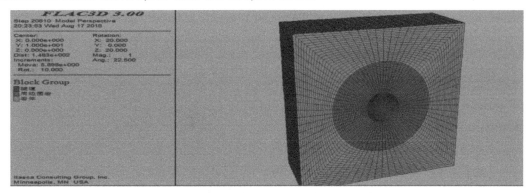

Figure 5.12   Tunnel model

(2) Stress boundary conditions

The calculated buried depth of the tunnel is 1,200 m, which belongs to a deeply buried tunnel, so that the tectonic stress is relatively obvious. After the geostress test, the maximum principal stress direction is NW 65°–85°. Combined with the tunnel axis direction, the maximum stress in the normal direction of the tunnel can be determined to be 44.5 MPa. On-site geostress conditions are achieved by applying stress directly inside the model.

Boundary conditions: to fix the speed of each boundary and pot constraints on the left and right (X direction), front and back (Y direction), and upper and lower boundaries of the model.

The initial stress field is shown in Figure 5.13.

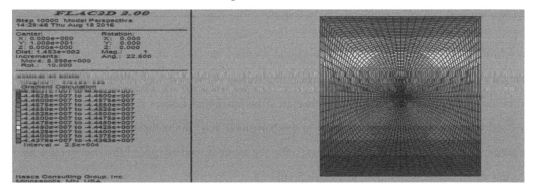

Figure 5.13   The initial stress field

4) Simulation calculation of dynamic load application

The dynamic calculation is based on static analysis. The single-hole simulation calculation is performed by using ANSYS/ LS-DYNA. Then the post-processing calculation results are listed in an Excel file, and applied as a load to the FLAC 3D software, where the overlay simulation calculation of porous blasting stress is performed (Figure 5.14). After calculation, we can judge whether the maximum principal stress of the tunnel is greatly reduced (Figure 5.15).

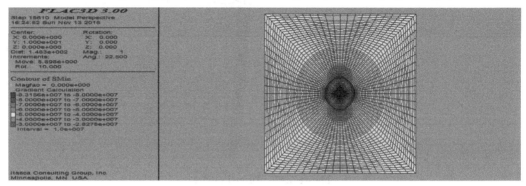

Figure 5.14 Maximum principal stress of the tunnel before de-stress blasting

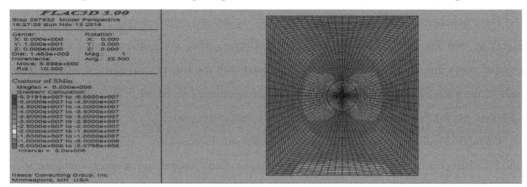

Figure 5.15 Maximum principal stress of the tunnel after de-stress blasting

5) Data processing

In the simulation analysis of the numerical results, we focus on the stress in the simulation results. In calculation, the stress in the tunnel is usually calculated, according to the elastic stress in the deep buried circular tunnel. In this state, the horizontal load is symmetrical to the vertical axis and the vertical load is symmetrical to the horizontal axis. Since the horizontal principal stress is greater than the vertical principal stress; take the vertical load as p_0 and the transverse load $\lambda \times p_0$, the stress diagram is shown in Figure 5.16.

Therefore, the total stress solution is:

$$\sigma_r = \frac{1}{2}(1+\lambda)p_0\left(1 - \frac{R_0^2}{r^2}\right) - \frac{1}{2}(1-\lambda)p_0\left(1 - 4\frac{R_0^2}{r^2} + 3\frac{R_0^2}{r^4}\right)\cos 2\theta \qquad (5.14)$$

$$\sigma_\theta = \frac{1}{2}(1+\lambda)p_0\left(1 + \frac{R_0^2}{r^2}\right) + \frac{1}{2}(1-\lambda)p_0\left(1 + 3\frac{R_0^2}{r^4}\right)\cos 2\theta \qquad (5.15)$$

Chapter 5 Optimization of Parameters for Pre-release De-stress Controlled Blasting Based on Numerical Simultations

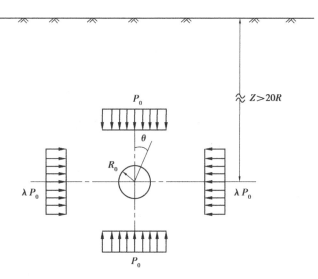

Figure 5.16 Tunnel load distribution when λ>1

$$\tau_{r\theta} = \frac{1}{2}(1-\lambda)p_0\left(1 + 2\frac{R_0^2}{r^2} - 3\frac{R_0^2}{r^4}\right)\sin 2\theta \qquad (5.16)$$

Based on the test results and actual test conditions of surrounding rocks, Linsheng Xu and Lansheng Wang obtain the following classifications of rockbursts:

$$\left.\begin{array}{ll} \sigma_\theta/\sigma_c < 0.3 & (\text{No rockburst activity}) \\ 0.3 \leqslant \sigma_\theta/\sigma_c < 0.5 & (\text{Slight rockburst activity}) \\ 0.5 \leqslant \sigma_\theta/\sigma_c < 0.7 & (\text{Medium rockburst activity}) \\ 0.7 \leqslant \sigma_\theta/\sigma_c & (\text{Intense rockburst activity}) \end{array}\right\} \qquad (5.17)$$

The maximum principal stress and the minimum principal stress in the tunnel are calculated according to the simulation, and the tangential stress σ_θ in the tunnel wall is calculated, according to the formula. Then based on the test, the uniaxial compressive strength of the andesite is 100 MPa. Therefore, the formula for calculating the discriminant coefficient is as follows:

$$K = \frac{\frac{1}{2}(1+\lambda)p_0\left(1 + \frac{R_0^2}{r^2}\right) + \frac{1}{2}(1-\lambda)p_0\left(1 + 3\frac{R_0^2}{r^4}\right)\cos 2\theta}{80 \times 10^6} \qquad (5.18)$$

5.3 Scheme for the De-stress Controlled Blasting

5.3.1 Overall plan

In designing the overall plan, two major numerical simulation contents in this chapter are taken into account: the numerical simulation of multi-factor coupling orthogonal experiment method for de-stress controlled blasting, and the numerical simulation of de-stress blasting for single factor changes.

De-stress controlled blasting is a complex process with multiple parameters interacting and interfering with each other. The blasting effect depends on these parameters and their combinations. The ANSYS/LS-DYNA is used to simulate the initiation process of de-stress controlled blasting. The numerical simulation parameters are derived from actual engineering and existing equipments. The specific blasting parameters include seven factors in total: the blasting method, the blasthole length, the blasthole diameter, the blasthole spacing, the blasthole uncoupled coefficient, the angles between the blasthole axis and the tunnel axis in the X and Y directions.

Even though there are only two levels in initiation modes, every other blasting parameter has 3 levels. If each level is combined to simulate the initiation process, a total number of numerical simulations are required, the number of tests is too large to manage. Therefore, in order to scientifically and reasonably simplify the number of numerical simulations, the "orthogonal experiment method" in mathematics is introduced.

The orthogonal experiment method is designed for the study of multi-factors and multi-levels tests. Based on orthogonality, it is tested by selecting some representative points from comprehensive experiments. These representative points have the characteristics of "uniformity, dispersion, neatness and comparability". The orthogonal experiment method is the main method for fractional factorial design. The orthogonal experiment method uses the normalized positive experimental test table designed following the principle of orthogonal Latin squares to reasonably combine the parameters to arrange numerical simulation, and process the obtained simulation results, with mathematical statistics methods, to obtain scientific and reliable conclusions. The numerical simulation in this book is obtained through a set of optimal blasting parameter combinations with the orthogonal experiment method.

The blasting parameters can affect the blasting effect through the changes of each combination of parameters, but the effect analysis of a single factor is not thorough. Therefore, in order to better understand and deeper explore the blasting effect of single factor on de-stress controlled blasting, the optimal parameter combination, obtained with the orthogonal experiment method, is used as the basic parameter group, and 5 relevant effect levels are set under each basic parameter, and some numerical simulations are used to study the effect of single factor on de-stress blasting.

5.3.2 Design for orthogonal experiment numerical simulation

According to the aim of the overall plan, the parameters of controlled blasting include seven factors in total as follows: the initiation mode, the hole length L, the hole diameter D, the uncoupled coefficient K, the hole spacing d, the angles θ_X, θ_Y between hole axis and tunnel axis in X and Y directions; the initiation method is in two levels: the forward initiation and the reverse initiation; the remaining seven factors are all in three levels, and the specific factor and level are shown in Table 5.6.

Table 5.6 List of factors and levels of orthogonal test

Factors Levels	Initiation way	Hole length L/m	Borehole diameter D/mm	Uncoupled coefficient K	Hole spacing d/cm	Angle in X direction $\theta_X/(°)$	Angle in Y direction $\theta_Y/(°)$
1	forward	4	35	1.0	70	30	30
2	reverse	5	50	1.5	100	45	45
3		8	90	2.0	200	60	60

It can be seen from Table 5.6 that the orthogonal experiment factor is a combination of 2×3^7. Combined with the orthogonal experiment design table, the orthogonal experiment table L_{18} is selected, and in combination with Table 5.1, the numerical simulation parameter combinations of de-stress controlled blasting are listed in Table 5.7.

Table 5.7 Design groups for orthogonal test of de-stress controlled blasting

Factors Levels	Initiation way	Hole length L/m	Borehole diameter D/mm	Uncoupled coefficient K	Hole spacing d/cm	Angle in X direction $\theta_X/(°)$	Angle in Y direction $\theta_Y/(°)$
1		4	35	1	70	30	30
2		4	50	1.5	100	45	45
3		4	90	2	200	60	60
4	forward	5	35	1	100	45	60
5		5	50	1.5	200	60	30
6		5	90	2	70	30	45
7		8	35	1.5	70	60	45

Continued

Factors Levels	Initiation way	Hole length L/m	Borehole diameter D/mm	Uncoupled coefficient K	Hole spacing d/cm	Angle in X direction θ_X/(°)	Angle in Y direction θ_Y/(°)
8	reverse	8	50	2	100	30	60
9		8	90	1	200	45	30
10		4	35	2	200	45	45
11		4	50	1	70	60	60
12		4	90	1.5	100	30	30
13		5	35	1.5	200	30	60
14		5	50	2	70	45	30
15		5	90	1	100	60	45
16		8	35	2	100	60	30
17		8	50	1	200	30	45
18		8	90	1.5	70	45	60

Modeling and analysis are performed according to the numerical simulation groups designed by the orthogonal experiment. Each group of simulations will have corresponding results, including the stress and strain, the velocity, the acceleration on the particle and the element. Data extraction is performed, according to the needs of modeling and result analyzing of ANSYS/ LS-DYNA and FLAC 3D. After extracting the results, combined with the features of the orthogonal experiment, a "range analysis" is performed, and then the final analysis results can be reached.

5.3.3 Design for single factor numerical simulation

Based on the overall plan and the result analysis of the orthogonal experiment in 5.3.1, the optimal parameter combination, obtained by the orthogonal experiment, is used as the basic data for the analysis of single factor de-stress effect. The de-stress effect of single factor will be analyzed from the numerical simulation of changes in basic data. The single-factor numerical simulation design is shown in Table 5.8.

Table 5.8 Design for single factor numerical simulation

Factor	Basic data	Single factor				
Hole length/m	5	3	4	8	10	2.5

Continued

Factor	Basic data	Single factor				
Borehole diameter/mm	50	35	40	70	90	110
Uncoupled coefficient	2	1	1.2	1.3	1.5	3
Hole spacing/cm	70	20	40	50	80	100
Angle in X direction/(°)	45	15	30	50	60	75
Angle in Y direction/(°)	30	15	45	50	60	75

5.4 Result Analysis of Numerical Simulations

5.4.1 Result analysis of orthogonal experiment numerical simulation

Firstly, from the FLAC 3D numerical simulations, the stress of surrounding rocks after simulated de-stress blasting in 1 to 18 groups is extracted respectively. Then according to formulas (5.1) and (5.2), we can get the discrimination coefficient K of de-stressing orthogonal experiment, as shown in Table 5.9.

Table 5.9 Discriminant coefficient of de-stressing orthogonal experiment

Experimental group	1	2	3	4	5	6	7	8	9
De-stress discriminant coefficient	0.437,2	0.439,9	0.436,3	0.380,5	0.259,6	0.258	0.424,2	0.431,4	0.424,2
Experimental group	10	11	12	13	14	15	16	17	18
De-stress discriminant coefficient	0.437,2	0.435,8	0.436,2	0.259,6	0.258	0.391,6	0.393,8	0.393,8	0.420,7

Range analysis method, also called visual analysis method, has the advantages of simple calculation, intuitive image, and being easy to understand. It is the most commonly used method for the analysis of orthogonal test results, including two steps: calculation and judgment analysis. In calculation, the de-stress discrimination coefficient K is derived from Table 5.10, and the maximum and minimum values of each factor are differentiated. The extreme difference is used to judge and analyze the influence of the factor, and finally the optimal combination of de-stress controlled blasting parameters is determined.

Table 5.10 Range analysis of discriminant coefficient of simulated distress in each group

Factor \ Index	S_1	S_2	S_3
Hole length	1+2+3+10+11+12	4+5+6+13+14+15	7+8+9+16+17+18
Borehole diameter	1+4+7+10+13+16	2+5+8+11+14+17	3+6+9+12+15+18
Uncoupled coefficient	1+4+9+11+15+17	2+5+7+12+13+18	3+6+8+10+14+16
Hole spacing	1+6+7+11+14+18	2+4+8+12+15+16	3+5+9+10+13+17
Tilt angle-x	1+6+8+12+13+17	2+4+9+10+14+18	3+5+7+11+15+16
Tilt angle-y	1+5+9+12+14+16	2+6+7+10+15+17	3+4+8+11+13+18
Initiation way	1+2+3+4+5+6+7+8+9	10+11+12+13+14+15+16+17+18	

(1) Extreme difference analysis calculation

According to the principle of the extreme difference analysis, the factors of de-stress controlled blasting can be combined, with a certain method, to calculate the total indexes of all factors. For the specific combination of factors, see Table 5.10.

Combine the de-stress discrimination coefficient K of each group according to Table 5.10, and calculate the extreme difference value; see Table 5.11 for details.

Table 5.11 Influences of various factors on the distress discrimination coefficient

Factor \ Influence	S_1	S_2	S_3	Polar difference
Hole length	2.184,2	1.505,2	2.072,2	0.679,0
Borehole diameter	1.942,7	1.847,7	1.971,3	0.123,7
Uncoupled coefficient	2.051,4	1.865,8	1.844,5	0.206,9
Hole spacing	1.860,5	2.060,0	1.841,2	0.218,7
Tilt angle-x	1.845,8	1.966,0	1.949,9	0.104,1
Tilt angle-y	1.839,8	1.952,5	1.969,1	0.129,3
Initiation way	3.666,1	3.598,19		0.033,9

(2) Extreme difference analysis judgment

Use the de-stress discriminant coefficient K to draw a trend map of each influencing factor, and according to the rule that the smaller the de-stress discriminant coefficient, the better the de-stress effect, we get what in the trend graph of each influencing factor, the smaller the influence value is, and the more favorable it is to for de-stressing.

From the analysis of Figure 5.17, we can get the changing process of the blast hole length L from 4 m to 8 m. In the change from 4 m to 5 m, the effect of the blast hole length on the de-stress effect is getting smaller and smaller with a minimum impact value of 1.5 and a drop impact difference of 0.68; during the change from 5 m to 8 m, the effect of the length of the blast hole on the de-stress effect increases gradually, with a maximum impact value of 2.1 and a rise effect difference of 0.57. Therefore, from the perspective of de-stress effect, the choice of the blasthole length should be about 5 m.

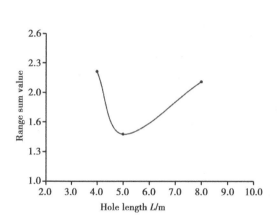

Figure 5.17 Trend graph of the influence of blasthole length L on the effect of de-stress blasting in orthogonal experiment

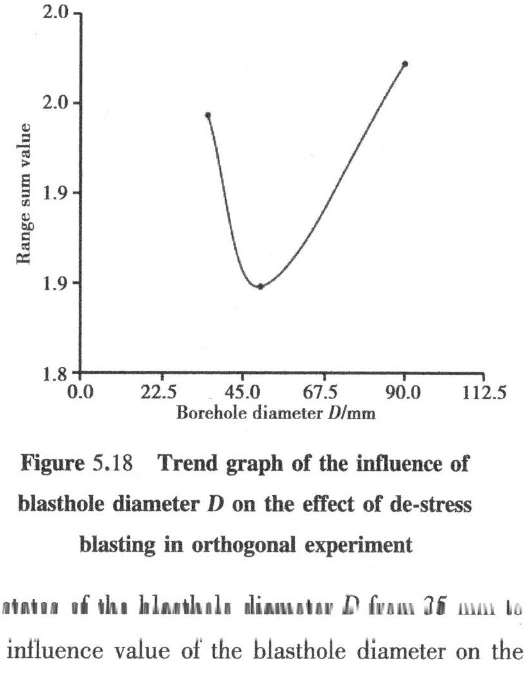

Figure 5.18 Trend graph of the influence of blasthole diameter D on the effect of de-stress blasting in orthogonal experiment

From Figure 5.18, we can get the changing status of the blasthole diameter D from 35 mm to 90 mm. At the stage from 35 mm to 50 mm, the influence value of the blasthole diameter on the de-stress effect is getting smaller and smaller, with a minimum of 1.84, and a decreasing influence difference of 0.1; during the changes from 50 mm to 90 mm, the influence value gradually increases again, reaching a maximum of 1.97, and the difference between the rising influence was 0.13; and the influence value of the blast hole at 90 mm is greater than that of 35 mm. Therefore, considering the best de-stress effect, the appropriate value of the blasthole diameter should be about 50 mm.

From the analysis of Figure 5.19: the influence value of the uncoupling coefficient in the orthogonal experiment of de-stress blasting effect is always reduced when the uncoupling coefficient K is changed from 1 to 2; and when K is changed from 1 to 1.5, the influence value decreases rapidly from 2.05 to 1.86, and the impact difference is 0.19; When K changes from 1.5 to 2, the influence value drops from 1.86 to 1.84, and the difference is 0.02. Compared with the former, the latter only fluctuates within a small range and can hardly affect the de-stress effect. Therefore, when the uncoupled coefficient K is 1.5, an ideal de-stress blasting effect can be achieved.

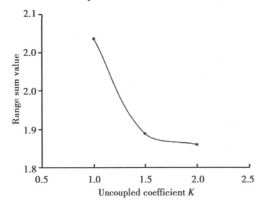

Figure 5.19 Trend diagram of the influence of uncoupling coefficient K on the effect of de-stress blasting in orthogonal experiment

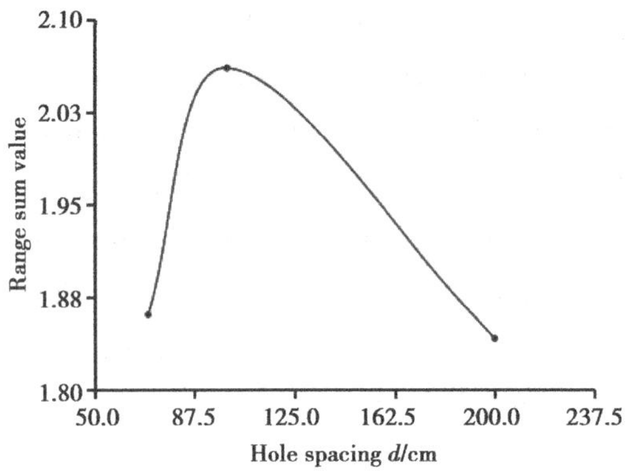

Figure 5.20 Trend diagram of the influence of blasthole spacing d on the effect of de-stress blasting in orthogonal experiment

From the analysis of Figure 5.20, we can get the changes of the blast hole distance d from 70 mm to 200 mm. When the distance d changes from 70 mm to 100 mm, the influence value on the de-stress blast effect increases from 1.86 to 2.06, and the increasing difference is 0.2. During the interval d changes from 100 mm to 200 mm, the value of the de-stress effect decreases from 2.06 to 1.84, and the drop difference is 0.22. Compared with the increasing difference, the relative rise difference is very close. In the orthogonal test of de-stress blasting effect, when the blast hole distance is about 100 mm, the de-stress blasting effect is not favorable. Therefore, from a comprehensive point of view, in order to obtain a good de-stress blasting effect, the appropriate interval of d is $d \leqslant 70$ mm.

From the analysis of Figure 5.21, we can get the changes of the angle between the blasthole axis and the model's X axis from 30° to 60°. Within the range from 30° to 45°, the influence value of the angle on the de-stress blast rises from 1.84 to 1.96, and the rising difference is 0.1. When the angle is from 45° to 60°, the influence value decreases from 1.96 to 1.94, and the decreasing difference is 0.02. Thus compared with the rising difference 0.1, when the angle is greater than 45°, the effect on de-stress blasting is smaller. Therefore, when the angle $\theta_X \leqslant 45°$, a better de-stress effect can be obtained.

From the analysis of Figure 5.22, we can get that, with the changes of the angle between the blasthole axis and the Y axis of the model from 30° to 60°, the influence of the angle θ_Y tends to increase. During the changes from 30° to 45°, the influence value increases from 1.83 to 1.95, and the rising difference is 0.12. During the changes from 45° to 60°, the influence value rises from 1.95 to 1.96, and the rising difference is 0.01. Compared with the former, the latter is in a relatively stable stage, which has almost no effect on the de-stress effect. Therefore, when the angle $\theta_Y \leqslant 45°$, a better blasting effect can be obtained.

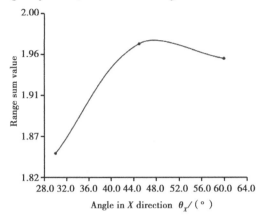

Figure 5.21 Trend diagram of the influence of angle θ_X in X direction on the effect of de-stress blasting in orthogonal experiment

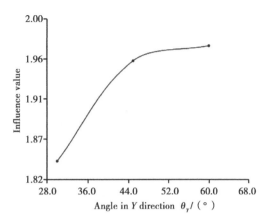

Figure 5.22 Trend diagram of the influence of angle θ_Y in X direction on the effect of de-stress blasting in orthogonal experiment

From the analysis of Figure 5.23, we can get that, as to the two ways to control initiation of de-stress blasting: forward and reverse, the effect of reverse initiation is better than the forward initiation. In this figure, when the initiation mode is positive, the impact value is 3.66; while in reverse initiation, the influence value is 3.59, and the influence difference is 0.07. Therefore, in this orthogonal test of de-stress effect, the reverse initiation is more conducive to the de-stress effect.

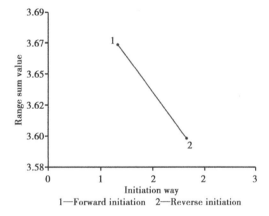

Figure 5.23 Trend diagram of the influence of initiation mode on the effect of de-stress blasting in orthogonal experiment

In summary, according to the discrimination coefficients of de-stress effect of each group in the orthogonal test, and the extreme difference values of the de-stress discrimination coefficients in Table 5.10, it can be concluded that the priority order of each parameter should be: the blasthole length, the blasthole spacing, the uncoupled coefficient, the Y-axis angle, the blasthole diameter, the X-axis angle, and the initiation mode. Therefore, according to the parameter selection priority, the results are shown in Table 5.12.

Table 5.12 **Optimization table of parameters for numerical simulation of de-stress blasting by orthogonal experiment**

Optimization of parameters for de-stress blasting	Parameter size
Hole length	5 m
Hole spacing	70 mm
Uncoupled coefficient	1.5
Dip θ_Y	30°
Borehole diameter	50 mm
Dip θ_X	30°
Initiation way	Reverse initiation

Then the values of the discriminant coefficients for the de-stress controlled blasting can be designed, according to each group's orthogonal test results in Table 5.9. Among them, the discriminant coefficients of groups 6, 7, 14 and 15 are all less than 0.3. Thus, according to the discriminant coefficients, there is no rock burst in their groups. However, in this numerical simulation results of the orthogonal experiment, it is found that, among the 4 groups of experiment with good de-stress effect, according to the priority of the affected parameters, the 14th group is finally selected as the priory group, as shown in Table 5.13.

Table 5.13 **Optimization table of final parameters for numerical simulation of de-stress blasting effect by orthogonal experiment**

Optimization of parameters for de-stress blasting	Parameter size
Hole length	5 m
Hole spacing	70 mm
Uncoupled coefficient	1.5
Dip θ_Y	30°
Borehole diameter	50 mm
Dip θ_X	45°
Initiation way	Reverse initiation

5.4.2 Result analysis of single-factor numerical simulation

According to the single-factor numerical simulation in Table 5.8, using ANSYS/LS-DYNA and FLAC 3D and analyzing the results of each group of numerical simulation, according to the above methods, we can obtain the results in Table 5.14.

Table 5.14 Numerical simulation results of single factor blasting

Factor		De-stress discriminant coefficient
Borehole diameter D/mm	30	0.348,5
	40	0.302,2
	70	0.254,5
	90	0.249,4
	100	0.247,9
Hole length L/m	2.5	0.452,3
	3.0	0.428,5
	4.0	0.326,6
	8.0	0.234,2
	10.0	0.218,1
Hole spacing d/cm	20	0.226,4
	40	0.243,6
	50	0.244,2
	80	0.318,6
	100	0.361,8
Uncoupled coefficient K	1.0	0.203,1
	1.2	0.219,7
	1.3	0.223,6
	1.5	0.231,4
	3	0.325,2
Angle in X direction θ_X/(°)	15	0.267,7
	30	0.261,2
	50	0.275,6
	60	0.301,2
	75	0.314,3

Continued

Factor		De-stress discriminant coefficient
Angle in Y direction $\theta_Y/(°)$	15	0.301,4
	30	0.269,3
	50	0.285,1
	60	0.291,4
	75	0.302,3

According to the numerical simulation results of each factor in the de-stress controlled blasting in Table 5.14, a series of influence trends can be drawn, as shown in Figure 5.24 – Figure 5.29. The specific charts and analysis are as follows:

From Figure 5.24, During the changes of the blasthole diameter D from 30 mm to 100 mm, with the increases of the blasthole diameter, the influence curve of the de-stress discrimination coefficient keeps a downward trend, which indicates that increasing the blasthole diameter is conducive to release stress. When the diameter of the blasthole is in the range of 30 mm$\leqslant D \leqslant$40 mm, the influence curve decreases significantly, and the de-stress discrimination coefficient decreases from 0.345,8 to 0.302,2. At this time, a slight rock burst might still occur, and the de-stress effect is not ideal. In the range of 40 mm$<D\leqslant$110 mm, the influence curve first gradually stabilizes in a relatively steep trend, and the de-stress discrimination coefficient decreases from 0.302,2 to 0.247,9 (basically less than 0.3), which shows that the de-stress effect is ideal. Thus for safety and reliability, it is more appropriate to control the diameter D of the blasthole in the range from 50 mm to 70 mm.

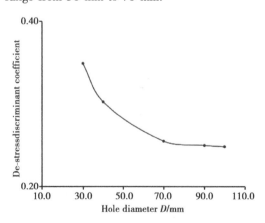

Figure 5.24 Trend diagram of the influence of blasthole diameter D on the effect of de-stress controlled blasting

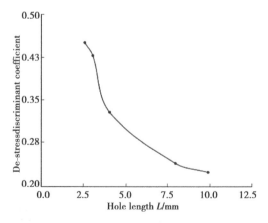

Figure 5.25 Trend diagram of the influence of blasthole length L on the effect of de-stress controlled blasting

From the analysis of Figure 5.25, we can get: in increasing the blasthole length L, the influence curve of de-stress discrimination coefficient keeps a downward trend, which indicates that increasing the blasthole length is favorable to release stress. When the blasthole length is in the range of 2.5 m $\leqslant L \leqslant$ 4 m, as L increases, the de-stress discrimination coefficient decreases significantly, and the value of the de-stress discrimination coefficient decreases from 0.452,3 to 0.326,6. Although it is reduced more significantly, the de-stress coefficients are all greater than 0.3, which indicates that the blasthole length in this interval is still considered to be detrimental to release stress. When the blasthole length is in the range of 4 m $< L \leqslant$ 10 m, the de-stress discrimination coefficient decreases from 0.326,6 to 0.218,1, and the downward trend is rapid and then gradually stabilizes. The de-stress discrimination coefficient in this section is less than 0.3 as a whole; but for the sake of safety, it is recommended that $L \geqslant 5$ m. Therefore, it is more appropriate to keep the blasthole length in the range of 5-8 m.

From the analysis of Figure 5.26, we find that with the increase of the blasthole distance d, the de-stress discrimination coefficient increases and the de-stress effect weakens. In the interval of 20 cm $\leqslant d \leqslant$ 80 cm, the trend diagram of de-stress discriminant coefficient changes smoothly, and the minimum value is increased from 0.226,4 to 0.318,6. But Overall, the de-stress discriminant coefficient is still less than 0.3, and a slight rock burst might occur near d=80 cm. In the interval of 80 cm $< d \leqslant$ 100 cm, the de-stress coefficient increases sharply, from 0.318,6 to 0.361,8, and the blasthole length values in the changed interval are all greater than 0.3, indicating rock bursts will occur. For the sake of safety, according to the analysis results, the suitable interval for blasthole spacing is: 20 cm $\leqslant d \leqslant$ 70 cm.

From the analysis of Figure 5.27: In increasing the uncoupled coefficient, the increase of the de-stress discrimination coefficient is more significant, but there are still differences in some intervals. When $1.0 \leqslant K \leqslant 1.5$, the de-stress discrimination coefficient increases from 0.203,1 to 0.231,4, and since the de-stress discrimination coefficient in this interval is less than 0.3, rock burst will not occur. When $1.5 \leqslant K \leqslant 3.0$, the influence curve of the de-stress discriminant coefficient increases significantly, from 0.231,4 to 0.325,2, then rockburst might occur near K = 3. As shown in the figure, the uncoupling coefficient has a basically linear relationship with the de-stress discrimination coefficient. Basically, only after $2.5 < K$, the de-stress discrimination coefficient is greater than 0.3, the rock burst might occur. Therefore, the suitable value of the uncoupled coefficient is: $1.0 < K < 2.5$.

From the analysis of Figure 5.28: As the angle between the axis of the blasthole and the axis of the tunnel in the X direction increases, the de-stress discrimination coefficient curve first decreases slightly, then rises significantly, and finally stabilizes. When the included angle θ_x is in the range of $15° \leqslant \theta_x \leqslant 30°$, the de-stress discrimination coefficient decreases from 0.267,7 to 0.261,2, thus the de-stress curve decreasing slightly. When the angle θ_x is in the range of $30° \leqslant$

$\theta_x \leqslant 60°$, the de-stress discrimination coefficient decreases from 0.267,7 to 0.301,2, thus the de-stress curve showing a significant upward trend, and rock burst occurring near $\theta_x = 60°$. When the included angle θ_x is in the range of $60° \leqslant \theta_x$, the de-stress curve tends to be stable; but at this time, the de-stress discriminant coefficients are all greater than 0.3, and rockburst will occur. For the sake of safety, the suitable angle θ_x between the blasthole axis and the tunnel axis in the X direction is: $15° \leqslant \theta_x \leqslant 50°$.

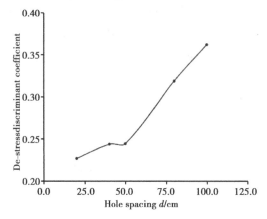

Figure 5.26 Trend diagram of the influence of blasthole distance d on the effect of de-stress controlled blasting

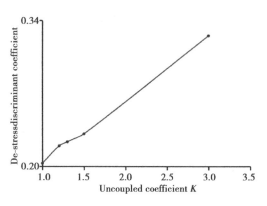

Figure 5.27 Trend diagram of the influence of uncoupling coefficient K on the effect of de-stress controlled blasting

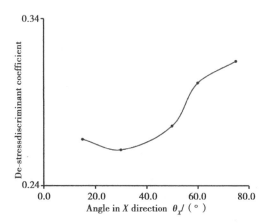

Figure 5.28 Trend diagram of the influence of angle θ_x in X direction on the effect of de-stress controlled blasting

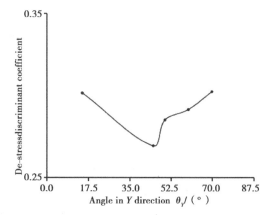

Figure 5.29 Trend diagram of the influence of angle θ_Y in Y direction on the effect of de-stress controlled blasting

From the analysis of Figure 5.29: With the changes of the angle between the blasthole axis and the tunnel axis in the Y direction, the curve of the de-stress coefficient first decreases linearly and significantly, then rises prominently, and finally rises steadily. When the included angle θ_Y is within the range of $15° \leqslant \theta_Y \leqslant 45°$, the de-stress coefficient decreases from 0.301,4 to 0.269,3, thus the de-stress curve decreasing linearly; When the included angle θ_Y is in the range of $45° \leqslant \theta_Y \leqslant 50°$, the de-stress coefficient decreases from 0.269,3 to 0.285,1, so that the de-stress curve almost jumps up, and the change is quitly obvious. Although the de-stress changes in the above two ranges are obvious, the de-stress discriminant coefficients are all less than 0.3, which indicates that rock burst will not occur. When the included angle θ_Y is within the range of $60° \leqslant \theta_Y < 70°$, the de-stress curve gradually stabilizes, and the rock burst will hardly occur; a rockburst might occur at $70° \leqslant \theta_Y$. Therefore, the appropriate angle θ_Y between the blasthole axis and the tunnel axis in the Y direction is: $15° \leqslant \theta_Y \leqslant 60°$.

In summary, based on the analysis of the influence trend of each single factor on the de-stress blasting, a suitable range of values is given for each single factor, specifically:

Comparing Table 5.13 with Table 5.15, we get that the results, obtained from the single factor analysis of the numerical simulation, are basically consistent with the results of the orthogonal experimental numerical simulation. It shows that the analysis of the de-stress blasting effect is relatively unified, and the variation rules of parameters in de-stress blasting under the single factor test and the orthogonal test are obtained.

Table 5.15 **Optimal range of parameters for de-stress controlled blasting**

Optimization of parameters for de-stress blasting	Optimal range
Borehole diameter D	50–70 mm
Hole length L	5–8 m
Hole spacing d	20–70 cm
Uncoupled coefficient K	1.0–2.5
Dip θ_Y	15°–50°
Dip θ_X	15°–60°
Initiation way	Reverse Initiation

5.5 Summary

In this chapter, numerical simulation software ANSYS/ LS-DYNA and FLAC 3D is used to simulate the de-stress blasting of surrounding rocks under high-geostress conditions. We, First of all, simulate the de-stress effects of various parameters, under the condition of mutual change, mutual influence and interaction, and then simulate the de-stress blasting situation under the change of single parameter, and finally analyze the influence rule of single factor on de-stress effects. The main work contents and results are as follows:

①The advantages of finite element software ANSYS and finite difference software FLAC are analyzed. And through the comparison and comprehensive utilization of the two software tools, we decide to use the ANSYS software to simulate the initiation process of de-stress blasting, and use the FLAC software to simulate the de-stress effect of blasting in surrounding rocks. For the needs of field test and numerical simulation, the parameters in software simulation are comprehensively analyzed and determined.

②According to the actual situation of de-stress blasting under high geostress, the process parameters of numerical simulation of de-stress blasting are analyzed and determined, which mainly include seven items, namely the blasthole diameter, the blasthole length, the blasthole spacing, the blasthole uncoupling coefficient, the initiation sequence, the angles between blasthole and tunnel axis in X direction and Y direction respectively. Firstly, we design the overall plan for the de-stress blasting under multi-parameter interaction conditions and, by using the orthogonal experiment method, determine the blasting scheme of de-stress blasting under single factor condition, and finally we make the simulation analysis of de-stress blasting.

③Based on the numerical simulation results of each experiment group, with the orthogonal experiment method combined with the evaluation method of the discriminant coefficient of de-stress blasting, we firstly determine the suitable value range of each parameter under the premise of mutual changes. Then, by comprehensively analyzing the on-site blasting situations and the numerical simulation results, we decide a group of optimal combinations of parameters under various interactions or influences: the blasthole length is 5 m, the blasthole distance is 70 mm, the uncoupling coefficient is 1.5, the blasthole diameter is 50 mm, the angle between blasthole and tunnel is 45° in X direction, and 30° in Y direction, and the initiation mode is reverse initiation.

④According to the analysis of the single factor of de-stress blasting, and combined with the evaluation of the discrimination coefficient of de-stress blasting, we determine the suitable range of each parameter under the change of single de-stress blasting parameters: the blasthole length is 5-8 m, the blasthole spacing is 20-70 cm, the uncoupling coefficient is 1.0-2.5 m, the blasthole diameter is 50-70 mm, the angle between blasthole and tunnel is 15°-50° in X direction, and 15°-60° in Y direction, and the initiation mode is reverse initiation.

Finally, through comprehensive analysis, it is found that the numerical simulation results of de-stress blasting parameters, by the orthogonal experiment and the single factor experiment, are relatively unified, which is in accordance with the optimization rule of de-stress blasting parameters.

Chapter 6 On-site Tests of Stress Release by Borehole De-stress Blasting

Through numerical simulation calculation and the orthogonal analysis, a set of optimized de-stress blasting parameters are obtained: the blasthole length is 5 m; the blasthole distance is 70 mm; the uncoupling coefficient is 1.5; the blasthole diameter is 50 mm; and the initiation mode is reverse initiation. In order to further verify whether the optimal combination of parameters is effective in practical engineering, on-site tests are carried out. Since the Erlang Mountain Tunnel of Yakang Expressway has been completed at this time and does not have the on-site test conditions, after some related surveys, we choose the Sangzhuling tunnel in the Linla section of the Sichuan-Tibet Railway as the test site.

6.1 General Engineering Situation of Test Site

The Sichuan-Tibet Railway under construction is a key national planning project. The La (Sa)-Lin(Zhi) section is mainly laid along the Yarlung Zangbo River. The Sangzhuling Section in Sangri County, Shannan City has typical disaster features, such as high geostress and frequent rock bursts, which has been reported by CCTV because of its rockbursts. After numerical simulations, a set of optimized de-stress controlled blasting parameters for strong rockburst have been determined, and its effectiveness and reliability need to be verified on the spot. Therefore, the high-stress section of Sangzhuling tunnel is selected for on-site test. After several months' preparation, with the strong support from the designer China Railway Eryuan Enginnoring Group. Co., Ltd and the construction unit China Railway No.5 Enginnering Group. Co., Ltd, the research team carried out a series of tests in the Sangzhuling tunnel in April and August in 2017, which has achieved the expected purposes.

The Sangzhuling tunnel, part of the LLZQ-5 project in the newly-built Lhasa to Linzhi section of the Sichuan-Tibet railway, is located at the entrance of the Sangga Gorge between Woka Station and Bayu Station. The entrance mileage of the tunnel is DK173+655, and the exit mileage is DK190+104. The total length is 16,449 m, of which the single line tunnel is 14,890 m, the station tunnel is 1,559 m, and the maximum buried depth of the tunnel is about 1,347 m. The surrounding rocks in the tunnel are mainly granite and some slightly metamorphic gneiss, whose maximal intensity may be up to 210 MPa. Due to the tectonic effect and the influence of large buried depth, the geological conditions of this section are relatively complicated, and it is an area highly prone to rockbursts. The main tunnel is designed with a light rockburst section of 10,061 m, a medium rockburst section of 3,671 m, and an auxiliary guide pit is designed with a light rockburst section of 1,770 m, and a medium rockburst section of 655 m. The intensity of actual rockbursts is much higher than the designed ones, and the medium and severe rockburst sections exceed 2,000 m.

6.1.1 Rockburst features of Sangzhuling tunnel

The rock bursts in Sangzhuling tunnel have the features of suddenness, concealment and violentness. During a rock burst, the rock pieces are ejected from the parent rock, and they are generally thick in the middle, thin in rim and irregularly flaky in shape. With the increase of the rockburst intensity, the volumes of the rock pieces increase to be block-like or plate-like. Generally speaking, in the intact rock layer, the damaged zone is thinner, while in the rock layer with a small number of fissures, the damaged zone is much thicker. Rock bursts in the tunnel usually manifest as splitting failure or shear failure. The splitting failure is generally crispy and loud in sound, not large in scale, and most are peeled off from the parent rock in the form of flakes or shell-shapes; the peeling-off of rocks happens simultaneously with the explosive sound. The shear failure is generally dull and low in sound, but larger in scale and accompanied by smoke-like powder ejection; the rock pieces, produced by the damage, generally lag behind the blasting sound for 20 minutes to one hour, before falling off the parent rock.

In terms of time effect, rockbursts in the Sangzhuling tunnel generally occur within 4–24 hours after blasting, and the number of rockbursts decreases with time (Figure 6.1). In terms of space, there are two peak locations of rockbursts, one of which is within the range of 2 m in the tunnel face, and decreases gradually; the next of which is within the range of 1.2 to 1.4 times the diameter of the hole from the tunnel face. There are also cases where rockbursts occur at more than 100 meters away from the tunnel face. But most rockbursts are likely to occur at the top or at the waist.

The rockburst is induced by artificial excavation, which is directly related to the excavation methods and supporting measures. It is easy to occur in the place where the light explosion effect is poor, or the stress is concentrated. A large number of projects show that there is a certain relationship between rockbursts and the buried depth of the tunnel, but it is not that the greater the buried depth, the more likely a rockburst occurs. The high geostress in this tunnel mainly comes from the tectonic stress caused by the geological tectonic activity, rather than the gravity stress by the buried depth.

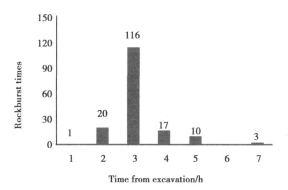

Figure 6.1　Time distribution map of rockbursts in inclined shaft of Sangzhuling tunnel

According to the scale and intensity, rockbursts can be divided into four types: the minor rockburst, the moderate rockburst, the strong rockburst and the severe rockburst. The minor rockburst is small in scale, shallow in pit (with a thickness of no less than 10 cm), and the length of rockburst crater is less than 10 m along the axis of the tunnel, and in sporadic distribution. In the moderate rockburst, the rockburst craters are usually triangular, arc-shaped or trapezoidal in shapes, large in scale, and tens of centimeters, even a maximum of 150 cm in pit depth, in piece distribution along the axis of the tunnel. The rockburst craters of the strong rockburst and the severe rockburst are distributed continuously, with a maximum depth generally no more than 4 m; most craters are about 1 m, with a length up to 20 m along the axis of the tunnel. The exfoliated rock blocks are large in size and huge in number, thus it may generate a large number of over-digging phenomena, making the cave shape irregular.

According to the relevant regulations of the "Technical Guide for the Construction of High-Speed Railway tunnels" and the "Code for Geological Survey of Hydropower Engineering", the macroscopic characteristics of rockbursts is described as follows Table 6.1.

Table 6.1　Macroscopic characteristics of rockburst intensity judgment

Macroscopic characteristics	Minor rockburst	Medium rockburst	Strong rockburst	Severe rockburst
Sound characteristics	Crackling sound, tearing sound	crisp popping sound	strong popping sound	violent dull sound
Motion characteristics	Loosening, peeling	Severe bursting and peeling	A large piece of film bursts, ejected or fell	Large bursts of continuous bursts and ejection of large rocks

107

Continued

Macroscopic characteristics	Minor rockburst	Medium rockburst	Strong rockburst	Severe rockburst
Aging characteristics	Sporadic intermittent bursts	Long duration with progressive progression to deep characteristics over time	Continuity and rapid expansion to deep surrounding rocks	Sudden and rapidly expands into the deep surrounding rocks
Engineering hazard	With minimal impact and proper safety measures, construction can proceed normally	If it has a certain impact, the measures of bolt-spary supports and anchoring the net should be hung up in time, otherwise it may develop to the deep part.	If it has a large impact, it should be supported by bolt-spary supports and anchoring the net in time.	If the project is seriously affected or even destroyed, corresponding special measures must be taken to prevent it.
Mechanical properties of bursting	Tension crack is the main failure.	Tension and shear damage coexist	Coexistence of shear failure	Coexistence of shear failure
Morphological characteristics of rock burst	Thin sheet, thin arc, thin lenticular	lenticular, ribbed	Edge plate, block, clintheriform	clintheriform, lump, or simple
Occurrence site	tunnel face, side wall and arch shoulder	Arch shoulder and arch waist	Mainly in the side walls and arches, which can affect the rest	Side walls and arches can spread to other parts
Fracture characteristics	Fresh shell shape	Shell shape, arc-shaped cavity, wedge shape	Large-scale arc-shaped cavity, wedge shape	Coexistence of large-scale arc-shaped cavity or wedge shape and shear failure
Influence depth	<1 m	1–2 m	1–2 m	>2 m

6.1.2 Rock bursts in Sangzhuling tunnel

The surrounding rocks of the Sangzhuling tunnel are mainly the granite Eocene solution mother-clava unit granite at the southern margin of the Gangdise plate, and the porphyritic hornblende biotite monophylete from the Tertiary Eocene Baidui unit Zone, fine-grained small porphyry biotite granite Tertiary Eocene animal husbandry unit, Tertiary Eocene Zhigegang unit, Tertiary Eocene animal husbandry unit and fine Cretaceous Menlang Unit fine-grained horndiorite fritite, hornblende biotite hornblende, biotite quartz pyrophyllite. Except for the entrance area of the tunnel, there are rockbursts of varying degrees in the other areas.

The No. 1 cross cave in Sangzhuling tunnel starts from H1DK0+580. After excavation, the rock mass bursts and peels off. During construction of the main tunnel, there are more rockbursts. After excavting the current face with a mileage of D1K176+463 (the Ⅲ grade surrounding rock, rock mass density 2.963 g/cm^3, strength 174 MPa, buried depth 450 m), there exists the splitting noise, part of the rock mass peeling off, whose area is less than 1 m^2.

The No. 2 cross cave in Sangzhuling tunnel starts from H2DK0+700. After excavation, the rock mass bursts and peels off. During construction of the main hole, after excavating the large and small mileage tunnel faces, there are obvious cracking sounds (lasting more than 12 to 24 hours, and there is still a muffled sound after concrete support), with peeling off of rock mass, whose area is generally about 2-4 m^2, some exceeding 8 m^2. And the depth of the blasting pits is about 50 cm, which mostly occurs near the arch waist. There are still rock spalling phenomena in the concrete support section, 15-20 m away from the tunnel face.

The small mileage of the tunnel face in the No. 2 cross cave is D1K179+781 (Ⅱ grade surrounding rock, rock mass density 2.892 g/cm^3, strength 170 MPa, buried depth 300 m). And the large mileage of the tunnel face in No. 2 cross cave is D1K181+251 (Ⅱ grade surrounding rock, rock mass density 2.72 g/cm^3, strength 152 MPa, buried depth 800 m).

In the inclined shaft of Sangzhuling tunnel, after the excavation of the tunnel face of XDK0+810, there exist some crack sounds and rock mass bursts. After XDK0+610, the crack sounds in the tunnel face are intensified, and the frequencies of rock mass bursts increase, with a blasting pit about 50 cm in depth, and about 2-4 m^2 in scope. The mileage from XDK0+280 to the current tunnel face XDK0+170 (Ⅱ grade surrounding rock, rock mass density 2.66 g/cm^3, strength 210 MPa, buried depth 800 m) exists strong and severe rockbursts, as shown in Figure 6.2: after excavation, the rock mass splits and cracks violently, and a large number of rock mass eject suddenly, which mainly occur in the arch, the arch waist or the cutting in the tunnel face, and the rockburst pit is more than 2 m in depth. At a distance of 20 m from the tunnel face, there are still some collapses of the side wall, caused by the ejections and cracks of rock mass, as shown in Figure 6.3.

In the main tunnel of the No. 4 Evacuation Exist DK189 + 485 to DK189 + 900, after excavating the tunnel face, there are crack sounds and rock mass collapses. The depth of the crater is generally about 30 cm and the scope is about 2 m². In the section from DK189 + 365 to DK189+020 in the single line section of the exit flat guide area, there are crack sounds and the rock blocks peel off after excavation. From DK189+020 to the current tunnel face D1K188+856 (II grade surrounding rock, rock mass density 2.936 g/cm³, strength 145 MPa, buried depth 1,350 m), violent crack sounds occur after on-site excavation, with rock mass cracking and ejecting, which is generally about 2 m² in scope, and the crater depth is 50 cm.

Figure 6.2 Strong rockburst in inclined shaft of Sangzhuling tunnel (once reported by CCTV)

Figure 6.3 Sidewall collapse caused by rockburst in inclined shaft of Sangzhuling tunnel

Based on the above situations, the test site was selected at the high stress section of the main tunnel, near the inclined shaft where a severe rock burst has occurred. In order to reduce the influence of excavation on stress redistribution, a refuge recess reserved on one side, is selected as the test site for stress-testing and de-stress blasting, and the stake number is DK186+020.94 in the main tunnel (the center position of stress test hole), where exist grade II surrounding rocks with a density of 2.66 g/cm³, a strength of 143.8 MPa and a buried depth of 850 m. The selection of this site also avoids the interaction between the test site and the excavation of the main cave.

6.2　Blasting Scheme Design

The test site is the refuge recess on one side of the main tunnel of Sangzhuling tunnel. The take mark is DK186+020.94, and the tunnel face is 8.7 m away from the side wall of the main tunnel. The pits, where the rocks are peeling off by rockbursts, can be seen clearly on the tunnel face.

According to the analysis of the single factor of de-stress blasting in the conclusion of chapter 5, and combined with the evaluation of the discriminant coefficient of de-stress blasting, the suitable value range of each parameter is determined (the blasthole length is 5-8 m, the blasthole distance is 20-70 cm, the uncoupling coefficient is 1.0-2.5, the blasthole diameter is 50-70 mm, the initiation mode is reverse initiation). The drilling and blasting parameters of the on-site test is designed as follows: the stress test hole is arranged at the approximate geometric center of the tunnel face, and the test hole is 50 mm in diameter and 5-8 m in depth (as long as a complete core can be obtained and the accurate stress data can be measured). On the simulated excavation contour, 7 pre-release de-stress blast holes are arranged in a circular shape, parallel to the axis of the hole, with a single hole depth of 6 m, a diameter of 50 mm, and a hole distance of 0.4 m, the uncoupling factor 1.0, and the reverse initiation method. It is blasted in the hole, and cracks are generated around the blasthole, thus forming an artificial crushing zone, with a certain safety distance and thickness, from the cave wall. The parameters of this test are shown in Table 6.2.

Table 6.2 Blasting parameter table

Number of blastholes	7
Blasthole depth	5 m
Borehole diameter	50 mm
Uncoupled coefficient	1.0
Initiation way	Reverse initiation
Single hole charge	1.2 kg
Distance between blast hole and stress test hole	1 m

The layout and side view of de-stress blast holes are shown in Figure 6.4 and Figure 6.5.

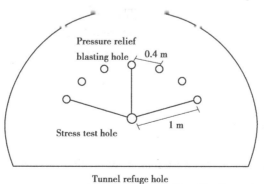

Figure 6.4 **Hole layout diagram of de-stress blasting**

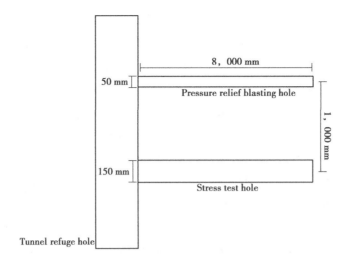

Figure 6.5   Side view of hole position in de-stress blasting

6.3   Test Equipment

The main equipment of this test include XY-2 geological drilling rig, test system of de-stress method, air-leg pneumatic rock drill, air compressor, etc., with the existing hydropower and pyrotechnic materials and the working conditions on site shown in Figure 6.6.

Figure 6.6   Field test

6.3.1   Stress test equipment

The stress test equipment is the Australian ES&S hollow inclusion strain gauge, whose model is CSIRO HID. This strain gauge has three strain flowers, and each strain flower has four strain cells (Figure 6.7), so the three-dimensional geostress can be measured by one drill hole.

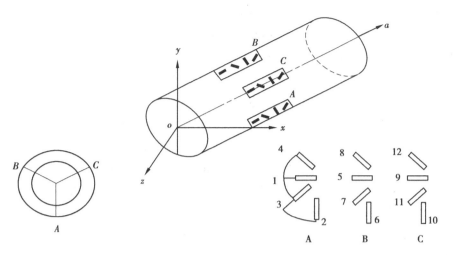

Figure 6.7 Schematic diagram of hollow inclusion strain gauge

6.3.2 De-stress blasting test equipment

The blasting test equipment and material are shown in Table 6.3.

Table 6.3 Blasting test equipment and material table

Equipment	Model	Quantity	Remarks
High energy detonator		2	—
Air hammer drill	T26	2	—
Core drilling rig	XY-2 Backdrilling Rig	1	—
Explosive	Emulsion explosive	8 kg	—
Air compressor	20 m³	1	—

6.4 Test Process

6.4.1 Experimental process of stress testing

①Drill the big hole. The XY-2 geological coring rig is adopted. After fixing the base, and adjusting the chain wheel of the drilling rig, we align the direction of the active drill pipe horizontally (slightly upward) to the hole position, and add a supporting rod on the back side to provide horizontal thrust for the drilling rig. A horizontal drill hole of 130–150 mm, is drilled on the measured tunnel wall, vertical to the rock surface, by a bit of 130 mm in diameter, and the depth is generally 3−5 times of the tunnel span, so as to ensure that the strain gauge is arranged in the stress area. The specific release position is determined by the condition of the borehole core. And the borehole is tilted slightly upward (about 3°−5° for drainage and easy to clean the borehole).

②Drill the tapered hole. After reaching the designed depth, we observe the core taken. If there are no cracks, we can start drilling the tapered hole. Smooth the bottom of the hole with a flat bit and hit the trumpet with a tapered bit, so as to facilitate the next step of drilling the concentric small hole, cleaning borehole and probing into the small hole smoothly.

③Drill the small hols. A concentric hole with a diameter of 36 mm is drilled from the bottom of the hole, and the depth of the small hole is 32−40 cm. After punching the small hole, clean it with water, and send the wipe head, soaked with acetone, into the small hole to scrub back and forth, to thoroughly remove the oil and other stolen materials in the small hole and dry them.

④Install the hollow core package. First, sand the outer cylindrical surface of the hollow core package with sandpaper. Then prepare the adhesive (two liquid materials, A and B) according to the proportion, pour an appropriate amount of adhesive into the cavity of the hollow core package to fix the pins and install the package on the director. Finally slowly feed a drill pipe into the large hole and note the length. Special attention should be given when the remaining length is about 5 m, to ensure that the package body can enter the hole intactly. When the front end enters the small hole for about 20 cm, the body of the package should be slowly pushed in; with the probe and the installer unhooked, the package body is successfully installed in the small hole.

⑤Initial strain data. After installing the package for about 20 hours, the epoxy resin is cured. The installer is carefully removed from the hole, and the electronic compass in the installer records the probe mounting angle. And the number shown is the mounting angle of the stress meter.

⑥Stress release. The thin-walled drill, used to make large hole in the first step, is applied to deepen the large hole, so that the core around the small hole can achieve stress release. The deformation or strain of the small hole, caused by stress release, is measured by a measurement system, including a test probe, and recorded by a recording instrument. According to the measured deformation or strain of the small hole, the geostress around the small hole can be obtained through a related formula.

⑦After the overcoring, the rock core and the strain gauge are taken out, the stress test ends. The drilling steps are shown in Figure 6.8 and Figure 6.9.

⑧The elastic modulus of the core is tested, by adding confining pressure with hydraulic Jack.

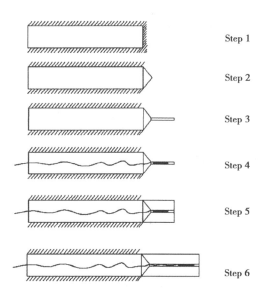

Figure 6.8 Recording process of stress release test data

Figure 6.9 Geostress test steps

6.4.2 Test process of de-stress blasting

①Drilling blast holes. The air leg pneumatic rock drill is used to drill blastholes, with the hole diameter of 50 mm, the depth of 5 m, the hole number of 7 and the blasthole distance of 1 m.

②Inspecting holes. The hole inspector, assigned by the blasting leader, carries out the hole-by-hole check-up, according to the blasting design. For those unqualified holes, timely remedial measures shall be taken, according to the opinions of engineers and technicians.

③Charging. Charging should be operated following the technical design requirements. In the process of charging, the charge height is measured with (wooden, bamboo) gun stick or gun rope, so as to avoid blocking the holes in the lower part. The charging capacity is 1.2 kg, i.e. a total of 6 rolls of 200 g cartridge, which are continuously installed at the bottom of the hole, and the hole is closed with gun mud. There is an air gap from the top of the explosive to the blockage of the orifice.

④Stuffing. After the charging is completed, special clay (yellow mud) is used to fill the blast hole orifice. Stones and flammable materials should not be used for filling. The blocking length is generally 20 cm.

⑤Detonating. Detonate at the same time. The initiation work shall be performed by an experienced blaster, designated by the blasting leader. Before the initiation, a detailed inspection of the detonator and other equipment is performed, to ensure a safe initiation. The detonator implemented the detonation, after receiving the command of the alert person and confirming that it is safe and reliable.

⑥Post-blast inspecting. The post-blast inspection is carried out by the blasting leader and experienced blasters, and the inspection should be carried out, after the blast smoke and dust have dispersed.

⑦After blasting, repeat steps (2) to (7) in 6.4.1.

6.5 Data Processing and Test Results

6.5.1 Experimental data

Before drilling the blast hole, the stress test hole is made with the XY-2 rotary drilling rig, in a depth of 5 m. The stress is tested by installing the hollow core package, and the original data of the pre-explosion stress are as follows: (The data is automatically generated by the instrument, a total of 9 groups, only one group used as example.)

$SPHC1,+0370,+1731,+0709,+0485,+1100,+0730,+0729,+0081,+0293,+0650,+0159,+0221,+21.0,00011*3E

$SPHC1,+0367,+1734,+0707,+0483,+1100,+0729,+0730,+0082,+0294,+0656,+0159,+0219,+21.0,00014*39

Null Offsets

$SPHC1,-0002,+0005,+0001,+0005,+0002,-0003,+0000,+0001,+0001,+0000,-0001,+0000,+21.0,00017*3C

Report Interval = 15 Sec

Report Interval = 1 Min-(60 Sec)

$SPHC1,-0004,+0006,-0005,+0004,-0002,-0007,-0004,+0006,+0004,-0003,-0003,-0004,+21.0,00080*33

$SPHC1,-0011,+0004,-0005,+0006,-0003,-0005,-0007,+0008,+0007,-0006,+0001,-0002,+21.0,00140*30

$SPHC1,-0014,+0005,-0007,+0003,-0003,-0005,-0005,+0009,+0004,-0004,-0004,-0004,+21.0,00200*33

$SPHC1,-0017,+0004,-0008,+0006,-0004,-0006,-0007,+0008,+0005,-0003,-0004,-0003,+21.0,00260*3B

$SPHC1,-0017,+0005,-0007,+0008,-0005,-0006,-0007,+0002,+0006,-0006,-0005,-0003,+21.0,00320*32

$SPHC1,-0019,+0007,-0008,+0009,-0002,-0006,-0010,+0004,+0005,-0005,-0006,-0006,+21.0,00380*3B

$SPHC1,-0022,+0006,-0011,+0007,+0000,-0008,-0006,+0005,+0005,-0007,
-0006,-0004,+21.0,00440*33

$SPHC1,-0023,+0004,-0011,+0006,-0001,-0010,-0004,+0002,+0004,-0009,
-0008,-0009,+21.0,00500*33

$SPHC1,-0023,+0002,-0010,+0008,-0001,-0011,-0007,+0003,+0006,-0009,
-0010,-0008,+21.0,00560*35

$SPHC1,-0026,+0004,-0010,+0004,-0002,-0010,-0007,+0000,+0003,-0010,
-0009,-0009,+21.0,00620*38

$SPHC1,-0024,+0005,-0013,+0007,-0001,-0011,-0007,-0001,+0007,-0006,
-0012,-0008,+21.0,00680*3C

$SPHC1,-0027,+0006,-0014,+0003,+0001,-0013,-0007,+0000,+0004,-0012,
-0007,-0008,+21.0,00740*33

$SPHC1,-0028,+0004,-0015,+0004,-0002,-0014,-0008,-0001,+0002,-0013,
-0012,-0011,+21.0,00800*32

$SPHC1,-0028,+0002,-0016,+0004,-0002,-0013,-0008,+0000,+0000,-0013,
-0009,-0009,+21.0,00860*30

$SPHC1,-0029,+0003,-0015,+0001,-0001,-0013,-0006,-0001,+0002,-0015,
-0009,-0011,+21.0,00920*34

$SPHC1,-0029,+0003,-0017,+0003,+0000,-0013,-0009,-0001,+0004,-0020,
-0012,-0012,+21.0,00980*3F

$SPHC1,-0029,+0003,-0015,+0002,+0000,-0014,-0006,-0003,+0000,-0014,
-0013,-0008,+21.0,01040*3B

$SPHC1,-0029,+0002,-0018,+0001,-0001,-0012,-0006,-0002,+0002,-0017,
-0010,-0012,+21.0,01100*38

$SPHC1,-0030,+0002,-0018,+0003,-0002,-0014,-0006,-0003,+0004,-0014,
-0015,-0011,+21.0,01160*33

$SPHC1,-0029,+0002,-0017,+0007,+0005,-0016,+0002,-0006,+0001,-0015,
-0017,-0017,+21.0,01220*33

Report Interval = 5 Min-(300 Sec)

Report Interval = 15 Min-(900 Sec)

Report Interval = No Report

Report Interval = 3 Sec

Report Interval = 15 Sec

$SPHC1,-0032,+0001,-0016,-0003,+0002,-0007,-0009,-0004,+0009,-0012,
-0002,-0007,+21.0,01290*30

Report Interval = 1 Min-(60 Sec)

Report Interval = 5 Min-(300 Sec)

Report Interval = 15 Min-(900 Sec)

Report Interval = No Report

Report Interval = 3 Sec

Report Interval = 15 Sec

Report Interval = 1 Min-(60 Sec)

$SPHC1,-0026,+0004,-0018,+0002,+0000,-0006,-0012,+0000,+0001,-0017,-0002,-0008,+21.0,01412*3C

$SPHC1,-0027,+0003,-0018,+0002,-0002,-0007,-0010,+0000,+0004,-0018,-0003,-0008,+21.0,01472*30

$SPHC1,-0030,+0002,-0018,+0001,+0002,-0008,-0012,-0001,+0000,-0018,-0001,-0006,+21.0,01532*35

$SPHC1,-0030,+0004,-0017,+0003,+0001,-0005,-0013,-0002,+0002,-0019,-0002,-0006,+21.0,01592*38

……

After the initial stress test is completed, the blasthole drilling is performed immediately, by using a wind-powered rock drill, with a total of 7 holes and the depth of 5 m. Charge and ignite the explosive. The blasting parameters are shown in Table 6.2. After the blasting, we continue to test the stress in holes. The stress test data at a depth of 5.5 m after 20 hours was as follows: (The data is automatically generated by the instrument, with a total of 9 groups, and we only choose one group as an example.)

Report Interval = 3 Sec

Report Interval = 15 Sec

Report Interval = 1 Min-(60 Sec)

$SPHC1,+0000,-0011,-0002,-0001,-0001,-0004,-0019,+0001,+0003,+0003,-0008,-0002,+20.9,00393*3B

Report Interval = 5 Min-(300 Sec)

Report Interval = 15 Min-(900 Sec)

Report Interval = No Report

Report Interval = 3 Sec

Report Interval = 15 Sec

Report Interval = 1 Min-(60 Sec)

$SPHC1,-0006,-0008,-0007,-0004,-0003,-0008,-0014,-0005,-0015,+0001,+0004,-0005,+20.9,00511*30

$SPHC1,-0009,-0009,-0008,-0006,-0004,-0006,-0019,-0001,-0010,+0006,
-0011,-0008,+20.9,00571*38

$SPHC1,-0008,-0009,-0005,-0003,-0004,-0007,-0017,-0001,-0012,+0000,
+0017,-0005,+20.9,00631*30

$SPHC1,-0006,-0010,-0007,-0008,-0006,-0005,-0015,-0002,-0014,+0003,
+0018,-0005,+20.9,00691*3E

$SPHC1,-0007,-0008,-0006,-0002,-0007,-0008,-0017,-0001,-0013,+0002,
-0005,-0006,+20.9,00751*32

$SPHC1,-0009,-0013,-0010,-0007,-0004,-0008,-0017,+0001,-0014,+0003,
-0005,-0007,+20.9,00811*3D

$SPHC1,-0007,-0007,-0007,-0006,-0006,-0008,-0016,-0001,-0017,+0000,
-0006,-0005,+20.9,00871*33

$SPHC1,-0010,-0009,-0007,-0010,-0003,-0010,-0020,+0000,-0013,+0005,
-0015,-0007,+20.9,00931*36

$SPHC1,-0007,-0010,-0006,-0009,-0007,-0008,-0021,+0001,-0012,+0003,
-0059,-0005,+20.9,00991*3B

$SPHC1,-0010,-0013,-0005,-0007,-0005,-0005,-0020,-0003,-0013,+0004,
+0000,-0004,+21.0,01051*38

$SPHC1,-0007,-0011,-0006,-0007,-0009,-0006,-0018,-0001,-0010,+0003,
+0000,+0000,+21.0,01111*3A

6.5.2 Data processing of stress test results

①In the stress releasing process, the relation formulas between the strain value, measured by the hollow core stress meter, and the geostress are as follows:

$$\varepsilon_\theta = \frac{1}{E}\{(\sigma_x + \sigma_y)k_1 + 2(1-v^2)[(\sigma_y - \sigma_x)\cos 2\theta - 2\tau_{xy}\sin 2\theta] - v\sigma_z k_1\} \quad (6.1)$$

$$\varepsilon_z = \frac{1}{E}[\sigma_z - v(\sigma_x + \sigma_y)] \quad (6.2)$$

$$\gamma_{\theta z} = \frac{4}{E}(1+v)(\tau_{yz}\cos\theta - \tau_{xy}\sin\theta)k_3 \quad (6.3)$$

ε_θ, ε_z and $\gamma_{\theta z}$ are the circumferential strain, the axial strain and the shear strain, measured by hollow core stress meter respectively. $\varepsilon \pm 45°$ is a strain value in a direction of $\pm 45°$ with the drilling axis, that is, the z-axis. k_1, k_2, k_3, k_4 are k coefficients, and k coefficient is not a constant, since it is related to the elastic modulus, the Poisson's ratio, the borehole diameter, the inner and outer diameter of hollow core passage and the radial direction of strain gauges.

a. calculation of k coefficients

$$k_1 = d_1(1 - v_1 v_2)\left[1 - 2v_1 + \frac{R_1^2}{\rho^2}\right] + v_1 v_2 \tag{6.4}$$

$$k_2 = (1 - v_1)d_2\rho^2 + d^3 + v_1 \frac{d_4}{\rho^2} + \frac{d^5}{\rho^4} \tag{6.5}$$

$$k_3 = d_6\left(1 + \frac{R_1^2}{\rho^2}\right) \tag{6.6}$$

$$k_4 = (v_2 - v_1)d_1\left(1 - 2v_1 + \frac{R_1^2}{\rho^2}\right)v_2 + \frac{v_1}{v_2} \tag{6.7}$$

b. The formulas for calculating the magnitude and directions of principal stress are as follows:

$$\sigma_1 = 2(-P/3)^{1/2}\cos(\omega/3) + J_1/3 \tag{6.8}$$

$$\sigma_2 = 2(-P/3)^{1/2}\cos[(\omega + 2\pi)/3] + J_1/3 \tag{6.9}$$

$$\sigma_3 = 2(-P/3)^{1/2}\cos[(\omega + 4\pi)/3] + J_1/3 \tag{6.10}$$

In the formula:

$$P = -J_1^2/3 + J_2$$

$$\omega = \arccos \frac{-\theta/2}{(-P^3/27)^{1/2}}$$

$$J_1 = \sigma_x + \sigma_y + \sigma_z$$

$$J_2 = \sigma_x \sigma_y + \sigma_y \sigma_z + \sigma_z \sigma_x - \tau_{xy}^2 - \tau_{yz}^2 - \tau_{zx}^2$$

②Processing of rockburst discrimination data.

The total stress solutions are:

$$\sigma_r = \frac{1}{2}(1 + \lambda)p_0\left(1 - \frac{R_0^2}{r^2}\right) - \frac{1}{2}(1 - \lambda)p_0\left(1 - 4\frac{R_0^2}{r^2} + 3\frac{R_0^2}{r^4}\right)\cos 2\theta \tag{6.11}$$

$$\sigma_\theta = \frac{1}{2}(1 + \lambda)p_0\left(1 + \frac{R_0^2}{r^2}\right) + \frac{1}{2}(1 - \lambda)p_0\left(1 + 3\frac{R_0^2}{r^4}\right)\cos 2\theta \tag{6.12}$$

$$\tau_{r\theta} = \frac{1}{2}(1 - \lambda)p_0\left(1 + 2\frac{R_0^2}{r^2} - 3\frac{R_0^2}{r^4}\right)\sin 2\theta \tag{6.13}$$

The most widely used rockburst classification in the project is as follows:

$$\begin{cases} \sigma_\theta/\sigma_c < 0.3 & \text{(No rockburst activity)} \\ 0.3 \leqslant \sigma_\theta/\sigma_c < 0.5 & \text{(Minor rockburst activity)} \\ 0.5 \leqslant \sigma_\theta/\sigma_c < 0.7 & \text{(Medium rockburst activity)} \\ 0.7 \leqslant \sigma_\theta/\sigma_c & \text{(Severe rockburst activity)} \end{cases}$$

After data processing, the rockburst discrimination coefficients are used to judge the control of rockbursts.

In addition, on the basis of previous research and combined with domestic engineering experience, Zhenyu Tao proposed that rockbursts would not occur when $\sigma_c/\sigma_1 > 14.5$. Rockburst will occur when $\sigma_c/\sigma_1 \leqslant 14.5$, and rockbursts are classified into 4 levels (Dexing Wu et al., 2005), as shown in Table 6.4. (σ_c, σ_1 are the uniaxial compressive strength and the maximum principal stress of the rock, respectively).

Table 6.4 Table of rockburst classification proposed by Zhenyu Tao

Rockburst classification	σ_c/σ_1	Description
I	>14.5	No rock burst and no acoustic emission
II	14.5-5.5	Low rockburst activity with slight acoustic emission
III	5.5-2.5	Medium rockburst activity with strong acoustic emission
IV	<2.5	High rock burst activity with strong cracking sound

6.5.3 Data processing results

Before blasting, the test point is 5.5 m inside the side wall of the refuge cave, which is actually 14.2 m away from the side wall of the main hole. After the above data processing and formula calculation, the geostress test calculation results are shown in Table 6.5.

Table 6.5 Results of geostress test before blasting

Test site	Lithology	Geostress parameters	Maximum principal stress σ_1	Intermediate principal stress σ_2	Minimum principal stress σ_3
Mileage 186+020.04	Granite	Magnitude/MPa	30.7	20.8	17.4
		Direction/(°)	250	145	25
		Dip/(°)	31	23	49
Description		The direction of the principal stress is the projection direction of the principal stress, expressed as a quadrant angle. The dip angle " - " indicates the depression angle, and the positive angle is the elevation angle.			

After blasting, the test point is 6.25 m inside the side wall of the refugecave, which is actually 14.95 m away from the side wall of the main hole. After the above data processing and formula calculation, the geostress test calculation results are shown in Table 6.6.

Table 6.6　Results of geostress test after blasting

Test site	Lithology	Geostress parameters	Maximum principal stress σ_1	Intermediate principal stress σ_2	Minimum principal stress σ_3
Mileage 186+020.94	Granite	Magnitude/MPa	11.5	4.2	3.8
		Direction/(°)	275	185	75
		Dip/(°)	7	3	83
Description	\multicolumn{5}{l|}{The direction of the principal stress is the projection direction of the principal stress, expressed as a quadrant angle. The dip angle " – " indicates the depression angle, and the positive angle is the elevation angle.}				

The sampling compressive strength σ_c, at the stress test point, is 143.8 MPa, and the pre-explosion discrimination coefficient σ_c/σ_1 is 4.68, which means this is a medium rockburst. After de-stress blasting, the pre-explosion discrimination coefficient is $\sigma_c/\sigma_1 = 12.5$, which means this is a minor rockburst.

The above data processing results show that:

①The maximum principal stress before blasting is 30.7 MPa. After blasting, the maximum principal stress is 11.5 MPa. Thus the maximum stress reduces by 63%, which indicates that blasting has an obvious effect on the reduction of the maximum principal stress.

②The direction of the maximum principal stress σ_1, at the test point before blasting, is 250°, and that of the maximum principal stress σ_1, after blasting is 275°. Thus the direction of the maximum principal stress is adjusted and changed before and after blasting, but the change value is small. The direction of the tunnel axis is about 188°, which intersects at a large angle. The dip angle of the maximum principal stress is 31° before blasting and 7° after blasting. The dip angle of the maximum principal stress varies, but they all belong to gentle inclination, intersecting with the small angle of the horizontal plane, and the principal plane is slightly inclined.

③According to the above-mentioned rockburst classification standard, it belongs to medium rockburst before blasting, and after de-stress blasting, it reaches minor rockburst or even close to non-rockburst activity. Thus the relief effect of de-stress blasting is remarkable.

④The suitable range of parameters, simulated and optimized in de-stress blasting are as follows: the blasthole length is 580 m, the blasthole distance is 20 cm, the uncoupling coefficient is 1.0 mm, the blasthole diameter is 50 mm, the initiation mode is reverse initiation and so on. The experimental results show that the combination of parameters is scientific and reasonable, and it has provided an effective solution to de-stress blasting.

6.6 Comparative Analysis

The on-site test results show that the combination of de-stress blasting parameters optimized by numerical simulation is reasonable and effective, thus the correctness of the numerical simulation results is verified in terms of practice. Are the derived formulas credible? This book also makes the following verification:

First of all, in order to facilitate calculation and comparison, we carry out the coordinate projection conversion. The stress test results in Tables 6.5 and 6.6 are projected to the three-dimensional coordinate system, composed of the tunnel axis direction (X), the horizontal direction (Y) perpendicular to the tunnel axis, and the vertical direction (Z) perpendicular to the tunnel axis. The azimuth of the tunnel axis, measured on site is 280°. The calculated results of the projection are shown in Table 6.7.

Table 6.7 Table of geostress results before and after blasting after projection treatment

Test site	Lithology	Geostress parameter	Stress σ_X	Stress σ_Y	Stress σ_Z
Mileage 186+020.94	Granite	Pre-explosion projection value/MPa	20.269	27.530	21.401
		Post-explosion projection value/MPa	4.671	10.904	3.953
Description	Tunnel axis direction (X), horizontal direction (Y) perpendicular to the tunnel axis, vertical direction (Z) perpendicular to the tunnel axis. The azimuth of the tunnel axis is 280°				

Under the three-dimensional condition, the residual stress formulas of de-stress blasting are as follows:

$$\begin{cases} \sigma_{VR} = \sigma_V - k\rho_r C_p Q^\alpha (r_1^{-\alpha} \cos \delta_1 + r_2^{-\alpha} \cos \delta_2 + \cdots + r_n^{-\alpha} \cos \delta_n) \\ \sigma_{ZR} = \sigma_Z - k\rho_r C_p Q^\alpha (r_1^{-\alpha} \cos \beta_1 + r_2^{-\alpha} \cos \beta_2 + \cdots + r_n^{-\alpha} \cos \beta_n) \\ \sigma_{HR} = \sigma_H - k\rho_r C_p Q^\alpha (r_1^{-\alpha} \cos \gamma_1 + r_2^{-\alpha} \cos \gamma_2 + \cdots + r_n^{-\alpha} \cos \gamma_n) \end{cases} \quad (6.14)$$

Where σ_{VR} is X axial residual stress value at de-stress point, σ_{ZR} is Z axial residual stress value at de-stress point, σ_{HR} is Y axial residual stress value at de-stress point, α is the stress decay index, calculated to be 1.7, C_p is velocity of elastic wave in rock, the wave speed is 6,500 m/s, σ_V is X axial geostress value, 20.269 MPa, σ_Z is Z axial geostress value, 27.530 MPa, σ_H is Y axial geostress value, 21.401 MPa, $r_1, r_2, \cdots, r_n$ are the distance between the de-stress point and the center of each pharmacy, and $r_n = \sqrt{(x-x_n)^2 + (y-y_n)^2 + (z-z_n)^2}$, $\delta_1, \delta_2, \cdots, \delta_n, \beta_1,$

$\beta_2, \cdots, \beta_n, \gamma_1, \gamma_2, \cdots, \gamma_n$ is the angle between the blasting stress of each blasthole and the X-axis, Y-axis and Z-axis at the de-stress point, and

$$\begin{cases} \cos \delta_n = \dfrac{|x - x_n|}{\sqrt{(x - x_n)^2 + (y - y_n)^2 + (z - z_n)^2}} \\ \cos \beta_n = \dfrac{|y - y_n|}{\sqrt{(x - x_n)^2 + (y - y_n)^2 + (z - z_n)^2}} \\ \cos \gamma_n = \dfrac{|z - z_n|}{\sqrt{(x - x_n)^2 + (y - y_n)^2 + (z - z_n)^2}} \end{cases}$$

ρ_r is density of rock, 2,680 kg/m³, k is coefficient related to rock and process parameters, 0.3, and Q is charge quantity, 1.2 kg.

According to the on-site test data, the number of blast holes is 7, and the distance is 0.4 m. The blast holes are arranged in an arc, and the center of the arc is the test hole; the radius of the arc is 1 m. The position parameters, the charge quantity and the mechanical property parameters are all substituted into the formula.

$\sigma_{VR} = 20.269 - 0.3 \times 2680 \times 3800 \times 1.2^{1.7} \times (1.118^{-1.7} \times 0.392/1.118 + 1.118^{-1.7}$
$\times 0.784/1.118 \times 2 + 1.118^{-1.7} \times 1.086/1.118 \times 2) \times 10^{-6} = 7.039(\text{MPa})$

$\sigma_{HR} = 27.530 - 0.3 \times 2680 \times 6500 \times 1.2^{1.7} \times (1.118^{-1.7} \times 0.5/1.118 \times 7) \times 10^{-6}$
$= 9.129(\text{MPa})$

$\sigma_{ZR} = 21.401 - 0.3 \times 2680 \times 3800 \times 1.2^{1.7} \times (1.118^{-1.7} \times 1/1.118 + 1.118^{-1.7} \times$
$0.92/1.118 \times 2 + 1.118^{-1.7} \times 0.84/1.118 \times 2 + 1.118^{-1.7} \times 0.76/1.118 \times 2)$
$= 2.841(\text{MPa})$

After comparison, the three-dimensional stresses, projected to the X, Y and Z axes after de-stress blasting, are 4.671 MPa, 10.904 MPa and 3.953 MPa respectively, while the calculated results are 7.039 MPa, 9.129 MPa and 2.841 MPa. Thus the calculated values in all directions are very close to the measured values. Since the main purpose of the book is to reduce the maximum principal stress within the safe value of no rockburst. Although there may exist some minor errors, it does not affect the identification of de-stress effects, so the result is acceptable. According to the results, the residual stress calculation formulas, derived in this book, can be verified with the field test results, thus the formulas are correct and reliable.

The calculation results are substituted into the following criteria:

$$\begin{cases} \sigma_\theta/\sigma_c < 0.3 & (\text{No rockburst activity}) \\ 0.3 \leq \sigma_\theta/\sigma_c < 0.5 & (\text{Slight rockburst activity}) \\ 0.5 \leq \sigma_\theta/\sigma_c < 0.7 & (\text{Medium rockburst ctivity}) \\ 0.7 \leq \sigma_\theta/\sigma_c & (\text{Intense rockburst activity}) \end{cases}$$

We can obtain that the point, before the pressure-relieving blasting, belongs to the moderate-intensity rockburst activity; after the de-stress blasting with optimized parameters, the maximum stress is effectively released, to the state without rockburst activity.

6.7 Summary

In this chapter, the combination of parameters, obtained from numerical simulation in Chapter 5, is applied to the Sangzhuling tunnel. The results show that, by applying de-stress blasting in tunnels with moderate and severe rockbursts, the maximum principal stress can be reduced by about 60%, so the rockburst can be reduced to no rockburst or minor rockburst. In the test, the stress release method is used to test the change of geostress, before and after de-stress blasting, and the rockburst criteria proposed by Linsheng Xu and the Zhenyu Tao rockburst criteria are also used in this book. The on-site test verifies that the optimized parameter combination, generated by numerical simulation analysis, is effective.

At the same time, after de-stress blasting, through the residual stress calculation formulas, derived in this book, the difference between the calculated results and the measured values is 12%, which does not affect the classification of rockburst intensity, thus the formulas being acceptable. The ratio of residual stress to the compressive strength is less than 0.3, which belongs to no rockburst activity. This shows that the calculation results of the formulas are credible and consistent with the on-site test results.

Therefore, by comparing and analyzing of the calculation results of the formulas with the numerical simulations, we find that the derived formulas are reliable and these methods can be mutually verified.

Chapter 7 Application of De-stress Controlled Blasting in Double-shield TBM Tunnel

The above analysis shows that the de-stress controlled blasting method can be used to pre-release the geostress in tunnels, and the calculation formulas of residual stress are derived, and, after numerical simulation and on-site test verification, a combination of optimal blasting parameters are recommended. This parameter combination has proved to be feasible and effective, by the verification in relevant tunnels of Lasa-Lingzhi Railway, which is built in the mining method. Furthermore, due to the sporadic, violent, and dangerous characteristics of rockburst ejection, the fully enclosed construction in double-shield TBM method, can undoubtedly provide a good protection. Therefore, in recent years, the double-shield TBM construction has been applied more frequently in the high geostress tunnels. Though the solid shield in double-shield TBM can protect personnel and equipment, it is sometimes easy to cause machines to card, due to high geostress or strong rockburst, which might cause a huge loss, like the TBM equipment being buried. Therefore, it is necessary to release the high geostress in advance, so as to reduce stress in the double-shield TBM tunnels.

The double-shield TBM itself has strong thrust and strong torque, so it is not easy to be jammed in moderate and minor rockbursts. To determine in advance the exact section of possible severe rockburst in the tunnel, and to take anti-jamming measures, such as pre-releasing stress or adjusting tunneling parameters, are important in ensuring safe construction. Thus in order to achieve a safe construction, some high-risk rockburst sections should be determined in advance, by way of geological forecast and microseismic monitoring, etc.; and measures, such as pre-releasing stress or adjusting adaption of equipment, should also be taken, in those high-risk tunnel sections, and the optimization of support and structure should also be adopted in the follow-up construction.

7.1 Engineering Overview of Test Site

The test site is a tunnel in Tibet, which is a part of a rural road project in Lingzhi. The total length of the tunnel is 4,789 m, with the direction of entrance being N82°E and the direction of exit being S72°E. The tunnel turns at the stake marked K9+208 m. The entrance elevation is

3,547.02 m. In order to facilitate drainage, 0.4% of the longitudinal slope is uphill to the exit, with an elevation of 3,566.18 m. The inner diameter of the tunnel is 8.1 m, and the excavated diameter is 9.13 m, and the maximum buried depth is about 830 m. Among them, in the section with a length of 4,610 m between stakes marked K8+848−K13+458, the double-shield TBM is used in construction.

The area along the tunnel belongs to the glacial landforms of alpine canyons with strong denudation and ice erosion. The mountains there are steep, the valleys are more developed, and the terrain slope is deeper. The tunnel crosses a certain mountain at the pass, with an elevation of 4,220 m. In areas above 4,000 m, there are mainly some angular peaks and edge ridges, formed by glaciation; and in areas between 3,500 m to 4,000 m, there are the glacier movement zones with strong ice erosion. From Figure 7.1, The entire northern slope was formed by the cutting of two branch ditches, formed by the glacier erosion in the upper and lower streams, and the Pypa ditch, formed by the convergence of the branch ditches. The whole layout is in a "Y" shape. There is perennial water in the ditches, and at the bottom of the ditches scatter a large number of glacier erosion rubbles, of different sizes, mostly with sharp edges, piled by water transportation (Figure 7.2).

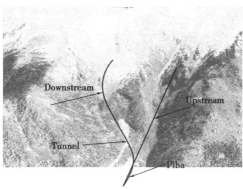

Figure 7.1 Land Features of south slope of the mountain

Figure 7.2 Land Features of North Slope of the mountain

From Figure 7.3, The south slope of the mountain is the early glacier formation area, which shows the third-order ice bucket landform, formed by the glacial movement cutting the river bed in early stage. The ice stages are composed of bedrocks, and between the ice stages are some platforms. The landforms are steep and gentle alternating step topography of the third stage ice bucket(Figure 7.4–Figure 7.6). The tunnel mainly penetrates the mountain on the right side of the south slope ice bucket, and the exit is located on the right side of the ice bucket.

Figure 7.3　land features of south slope

Figure 7.4　geomorphology of first-order ice bucket on the southern slope

Figure 7.5　The geomorphology of second-order ice bucket on the southern slope

Figure 7.6　The geomorphology of third-order ice bucket on the southern slope

In addition, there is an ice lake near the peaks on the north and south sides of the tunnel respectively, where the horizontal distance between the south ice lake and the tunnel is only 200 m, the elevation of the lake is about 4,300 m.

The tunnels are mainly the banded mixed gneiss (Figure 7.7) or the granitic gneiss (Figure 7.8) from the PT rock group (Pt2-3d) of the Nanga Bawa rock group in the Himalayan stratigraphic zone, with single rock lithology. In construction, through the direct observation from the cutter gap, the side window and the gaps at the back of the telescopic shield, it is found that the lithology of this section is mainly the banded mixed gneiss, with some granite gneiss partly developed. Affected by the different degrees of migmatization, some biotite hornblende gneiss

(Figure 7.9) and some feldspathic (Figure 7.10) and granitic banded (lump) veins are locally developed. The latent microfissures in some sarcoplasmic bodies are developed. Some local alteration zones are kaolinized, so the strength of rocks decreases after alteration, and they are easy to soften and disintegrate in water (Figure 7.11).

Figure 7.7　Compound gneiss

Figure 7.8　Granite gneiss

Figure 7.9　Hornblende gneiss

Figure 7.10　Sarcoplasmic body

Figure 7.11　The disintegration of the rock

The peripheral region of the project is in complex condition, but the structure along the tunnel is relatively simple, and there is no regional fault passing through. The structure is mainly represented as the Duoxiongla anticlines, the secondary fault fracture zones and the joint fractures, which are relatively developed in the project area.

During the geological re-examination of the PT anticline at this stage, a detailed investigation was carried out on the development of gneiss from the north slope, the estuary, and the south slope to the Lhasa-Golmud line. Through analysis, it is found that the PT mountain is an anticline structure with the anticline axis generally distributed in NNE direction. The dip angles of gneiss above the elevation of about 3,000 m are generally 25°–30°; below 2,500 m, the dip angles of gneiss are 60–80°. Combined with the topography of the south and north slopes of the mountain, it is found that the Duoxiongla anticline is a wide and gentle anticline, the axis of the anticline extends along the NNE direction, with tilts roughly to the north. The stratigraphic distribution of the two wings of the anticline is symmetrical, and the gneiss occurs slowly and the folds are gentle near the core of the anticline. The axis of the tunnel crosses the Duoxiongla anticline and intersects at a large angle with the direction of the anticline hub.

According to the geophysical profile tests of the imported deep borehole method and the surface EH4 magnetotelluric method, two large faults are revealed along the tunnel, of which F1 is exposed by coring in boreholes and F2 is inferred from the interpretation of EH4 test profiles. F1: the fault is exposed at the depth of 91.76 mm to 114.6 m at the entrance of the tunnel with a length of 22.84 m. The rock mass in this section is broken, and most of them are cataclastic rocks, silty rocks and a small amount of clastic rocks. The strength of the whole rock mass is very low, even can be crushed by hand, and some scratches can be found in the lower part of the fault. After analyzing the core from the in-hole TV and interpreting the gneiss occurrence, we can infer that the occurrence of the fault fracture zone is N50°–N55°W/NE∠42°, with a width of about 16 m. However, due to the thick cover layer at the presumed surface outcrop, no fault-related signs are found. F2: according to the results of the EH4 geophysical profile (Figure 7.12), there is an abnormal reflection segment, which is projected to the tunnel axis, combined with geological analysis, we find it has a width of about 220 m and appears at the stake marked K10+070–K10+290 in the tunnel. Inferred from the geophysical results, this section may be an exposed section with a certain scale of faults or a concentrated outcrop of several secondary faults. In addition, based on a comprehensive analysis of the regional geological conditions and multiple methods of drilling and geophysical exploration, it is believed that there may be more small-fault fracture zones along the tunnel, which mainly develop along the direction of gneiss.

During construction, the rock mass in the wall was directly observed through the gap behind the knife gap, side window and telescopic protective cover, and it was found that the gneiss had good integrity. The occurrence of gneiss is N10°–20°W/SW∠40°–45°; local faults and folds are developed. Some local joints and fissures are well developed, which are mainly manifested in the following four groups: ①Near SN/W(E)∠25–40°, extension 3–5 m, individual more than 5 m, spacing 0.5 – 2 m, closed, mostly filled with 1 – 2 mm thick crushed siltstone, and localized muddy. ② N70° – 75° W/SW ∠30° – 40°, extension 3 – 5, spacing 0.5 – 2 m, closed Micro-tensioned. ③ N70° – 80° E/SE ∠70° – 80°, extension 1 – 3 m, spacing 0.5 – 2 m, closed without filling. ④ N20 – 30° E/NW ∠70° – 80°, extension 1 – 3 m, spacing 0.5 – 1 m, closed without filling.

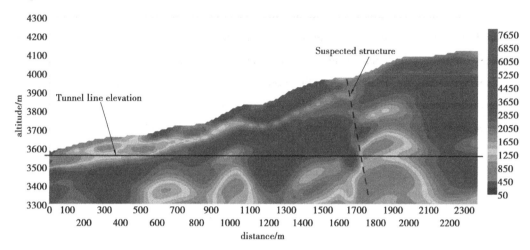

Figure 7.12 The result map of EH4 exploration and inversion in the north slope section of PM1 section

According to the excavation of the inlet section, we find the rocks in the inlet section are weakly weathered, the strong unloading depth is 30 m, and the weak unloading depth is 67 m. Through investigating the surface of the outlet section, we find that the unload of the surface rock mass in the outlet section is serious, the unloading fissures are generally open, and the unloading depth of the outlet section is larger than that of the inlet section. It is inferred that the strong unloading depth is about 40 – 60 m and the weak unloading depth is about 80 – 100 m. The weathering effect is relatively weak, so the surface rock mass is weakly weathered, and the weathering degree is similar to that of the inlet.

The rainfalls in this area are different between the north and south sides of the mountain. The annual precipitation of the north slope is about 800 mm, while on the top of the mountain and the south slope east of the mountain, the precipitation there, affected by the warm and humid air flow from the Indian Ocean, is evidently increased, and the annual average precipitation is more than 3,000 mm. The top of the mountain is covered with ice and snow all the year round, and the north and south sides of the mountain are abundant in surface water. There is perennial running water in the valley at the pass above the tunnel at a horizontal distance of 200 m from the tunnel, there is an ice lake there. There are two branch ditches of Paiba ditch on both sides of the entrance, where the elevation of water distribution is high, so the recharge condition of groundwater is sufficient. After analyzing the groundwater storage conditions, the characteristics of aquifer medium, the lithology and its combination, we judge the main types of groundwater are the bedrock fissure diving and loose accumulation of pore water, without finding obvious confined water phenomenon. The bedrock fissure water is mainly stored in the bedrock, which is related to the lithology, the degree of fissure development, and the weathering unload of the rock mass. The pore phreatic water is mainly stored in the Quaternary loose layers, such as alluvium, landslide deposit and ice deposit, which is greatly influenced by the seasonal variations and geomorphology. There are abundant groundwater recharges along the tunnel. At the shallow sections of the entrance and exit, and the cross-ditch section of the exit, there exist some infiltration channels, which may lead to local water gushing. However, the possibility of water gushing and inrushing is usually small in deep burial sections. Only when locally meeting the crush fracture zone or the fault fracture zone, some sudden water gushing might occur.

The surface water and the groundwater in tunnel area are generally colorless, tasteless and transparent. We get some ditch water from the Paiba ditch on the upstream side of the entrance, to make a chemical analysis of water quality. According to the Code for Highway Engineering Geological Investigation (JTJ C20-2011), we find the groundwater is slightly corrosive to concrete.

The tunnel is a buried mountain tunnel with a maximum depth of 830 m. In order to ascertain the geostress features of the area, a deep-borehole exploration is carried out at the entrance of the tunnel, and the hydrofracturing geostress test is performed in the borehole. At the same time, in the tunnel wall, the geostress test of the three-hole-intersection method is carried out. Based on these experimental results, combined with the empirical formulas, and the gravity features of stress, it is found that the tunnel sections with a depth greater than 400 m, are mostly the moderate and high geostress areas, and some minor rock burst may occur locally, and that when the buried depth is more than 600 m, the tunnel section belongs to the high stress area, which is prone to occur those phenomena of moderate rock bursts, sheeting side or loosening, and the corresponding length of the tunnel is about 2,200 m. When the buried depth is 830 m, the estimated maximum geostress is 30-35 MPa, and the corresponding stress ratio is 2-4, the section belongs to high stress area. Therefore, in this tunnel, the high geostress problems are prominent, and those bad geological phenomena, such as rock mass deformation, rock bursts, rock wall stripping, block falling and rock mass splitting etc., should be highly noticed, especially in those sections with large buried depth and huge intact rock mass.

The tunnel is constructed with the double-shield TBM. The Double-shield TBM, a new type of rock drilling machine, developed on the basis of open-type TBM and shield machine, is mainly composed of the host engine system and the rear supporting system. The host engine system includes the cutter head, the main bearing, the main drive motor, the main propulsion cylinder, the stabilizer, the anti-torque cylinder, the support boots, the auxiliary propulsion cylinder, the segment installer, the 1#belt conveyor, the front shield, the telescopic shield, the support shield, the tail shield and other components or shield bodies (Figure 7.13). The rear supporting system mainly includes the rear supporting belt conveyor system, the backfill grouting system, the bean gravel injection system, the segment crane and feeder, the crane and lifting equipment, the dust removal system, the secondary ventilation system, the compressed air system, the cooling water system, the water supply system, the drainage system, the hazardous gas monitoring and alarm system, the hydraulic systems, the slip systems, the main drive variable frequency control system, the power distribution systems, the lighting system, the communication system, etc., and it is also equipped with the VMT guidance system, the monitoring system and other auxiliary facilities.

Figure 7.13 Double-shield TBM mainframe

1—Cutter head 2—Slag collecting ring 3—Front shield 4—Main drive
5—Telescopic shield 6—Main propulsion cylinder 7—Torque cylinder
8—Equipment belt conveyor 9—Support shield 10—Supporting boot plat
11—Shield tail 12—Auxiliary propulsion cylinder 13—Segment installation machine

As the carrier of hob and side cutter, the cutter head also includes some heavy steel structures with functional parts, such as the cutter box, bucket, tooth, and slag collecting tanks. The main drive motor drives the main bearing to realize the rotation of cutter head. Under the action of the propulsion cylinder, the hob squeezes the broken rocks to achieve driving. As the heaviest part, the cutter head is generally divided into 2 or 5 pieces during transportation, and then assembled and welded after being transported to the site. The cutter head should have the function of expand digging 100 mm generally. If we want to achieve greater expand digging, the cutter head should be lifted to achieve normal excavation under the rock strength of 200 MPa. Some wear-resistant blocks are attached to the surface of the cutter head to prevent it from wearing. The hob includes single-edged and double-edged knives, which are divided into 19-inch and 17-inch according to diameter (some 20-inch knives are the ones with enlarged cutting ring). The maximum thrust of a 19-inch single blade is 315 kN, and the maximum thrust of a 17-inch blade is 267 kN, so the general thrust of the cutter head is: number of tools×315 kN (or

267 kN). The installation mode is usually backmounted, and the blade distance is set within the range of 80-100 mm. Adjusting the position of the edge knife can achieve 100 mm (diameter) expansion.

The double shield TBM has the advantages of high degree of automation, high construction efficiency, better cave-forming effect and construction safety, especially in those areas with a high risk of rockbursts. It can effectively prevent the occurrence of malignant accidents, including rockburst, so it has gradually become the first choice for the construction of high-stress tunnel. However, the cost of double-shield TBM is high, and it is easy to get stuck in the tunnel with high geostress, such as in the construction of Jinping II Hydropower Station, a large-scale TBM even got stuck. Therefore, it is urgent to find measures to prevent the double shield TBM from sticking in the tunnels.

7.2 Influencing Factors of Rockburst Stucks in this Tunnel

Rockburst is a dynamic failure phenomenon. Because of its suddenness and huge destructiveness, the local rock mass is relatively broken, due to the dense development and unfavorable combination of joints and fractures, resulting in tunnel instability. The advantage of double-shield TBM lies in its strong adaptability to all kinds of surrounding rocks. To the rock mass with certain self-stability, it has strong adaptability to pass through quickly and follow up the lining. However, for the fracture zone formed by large-scale faults (fractures), where the fracture zone is wider and the degree of fragmentation is severe, it is easy to cause the cutter disc to race or to produce a large amount of slags during TBM driving. When the slags are in large amount, it is difficult to rotate the disc and it may increase the current, which may cause the cutter head to jam. When a large-scale fault fracture zone appears in a high geostressed tunnel, the rock mass often undergoes stress-structural deformation and failure, and the shield body bears excessive stress, which causes shield to jam. At the same time, when the broken rock mass leads to a collapse, it also affects the construction of excavation and lining of segment.

At present, there is no unified standard and definition for the concept of high ground stress. It is generally believed that high ground stress refers to the initial stress, that is, when the maximum principal stress σ in the rock mass reaches more than 15% of the uniaxial compressive strength of the rock, the initial stress is considered high ground stress. Besides, in practical engineering, when the maximum principal stress of the initial stress is above 20-25 MPa, it is also considered as high geostress. The influence of geostress on rockburst can be understood from the following three aspects:

①The occurrence of rockburst is closely related to the elastic strain energy. Under the same geological background, some rocks have higher geostress, while others only have lower geostress. Usually, the elastic modulus of rocks with higher geostress is higher, and vice versa. Therefore, in high geostress area, the rock has more elastic strain energy, so the rockburst is most likely to occur, forming a rock fragmentation zone.

②The rock with high geostress is characteristic by brittle fracture, and rockburst is precisely the brittle failure process of rocks, which shows that the rocks in the high geostress area have the characteristics of a rockburst.

③When a tunnel is excavated in high geostress area, if the initial stress of the rock is disturbed and the stress around the tunnel is redistributed, the peak stress may reach 2-3 times of the initial stress. Due to the influence of concentrated stress, the stress in the surrounding rock often exceeds the critical stress of rockburst, resulting in rockbursts.

A large number of rockburst records show that almost all rockbursts occur in those elastic-brittle rock mass, with fresh and intact body, hard texture, high strength, dry without groundwater and large thickness of overlying rock layer. However, rockbursts seldom occur in those rock masses, with developed structural planes, large deformation, low strength and rich water. If the high geostress is regarded as the main external cause of rockburst, the lithology is then the main internal cause of rockburst. The influence of the lithology on rockburst is mainly reflected in the following aspects:

①The intensity of the rock. It is known from the practice engineering that, rockbursts are more prone to occur in the igneous rocks with single axis compressive strength exceeding 150 MPa, or in the sedimentary rocks exceeding 60 MPa. On the one hand, it shows that the type of rock affects the generation of rockburst. On the other hand, it shows that the rock intensity also affects the possibility of rockburst. Generally, the greater the compressive strength of rock, the bigger the probability of rock burst. This is because the stronger the rock is, the greater the ability to store elastic strain energy in the process of tectonic or shallow transformation. When the rock is destroyed, the remaining elastic strain energy after dissipation is converted into larger kinetic energy, which ejects or even throws rocks out. Under the same environmental conditions, the greater the intensity of the rock, the more possibility of rockburst, and the higher the intensity of rockburst, and vice versa.

②The integrity of the rock. The bright and intact rock with few primary fissures is similar to an elastic body, which can effectively store a large amount of elastic strain energy, so that less energy is consumed during rock failure, and the fractured rock mass can obtain enough kinetic energy to be ejected and thrown out, which may lead to the occurrence of rockburst. As for the rock with obvious fissures and joints, since its energy has been released in the process of formation, even if it undergoes the tectonism or strong shallow transformation in later stage, there is no conditions to store much energy because the rock mass is relatively broken, so rockburst will not occur. Some research data shows that if the integrity coefficient of rock is more than 0.75, it belongs to intact rock mass and is the condition for rockburst. According to some practical statistics, rockbursts occur in about 70% of intact and hard magmatic rocks, and only about 30% of rockbursts happen in those fresh and hard metamorphic and sedimentary rocks.

③The water content of the rock. In fact, most rockbursts occur in dry rocks, which can be explained by the role of water in rocks. The water-bearing rocks have high porosity, full of large number of well-developed joints and fractures, which, on the one hand, reduces the surface energy between rock particles, on the other hand, reduces the strength and elastic modulus of the rock, thus it leads to the softening of rock and the inability of storing and accumulating elastic strain energy, and then reduces the possibility of rockburst.

From some practical engineering projects, it is found that rockburst is easy to occur in the fold area, especially in the core area near the fold, where the tectonic stress is concentrated. Rockburst is not easy to occur in the broken rock mass in the fault zone, the fracture zone and the joint dense zone, while it is easy to occur in the intact rock mass close to the fault. Geological structure also has an obvious influence on rock burst. When studying the relationship between faults and rockbursts, it is necessary to study the mechanical properties of faults. Generally, in the footwall of the fault, where the rock mass is relatively complete and the stress is easy to concentrate, the intensity of rockburst is relatively high. Besides, rockburst is not easy to occur in caves with rich groundwater (the fault water, the fissure water, the karst water, etc.), but it is easy to occur in dry caves. Furthermore, it is necessary to pay attention to the relationship between high-head confined water and rockburst, since the confined water with high water head complicates the stress in the tunnel.

①The influence of engineering construction on rockburst. The influence of engineering construction on rockburst can be analyzed from two aspects: the excavation depth of the chamber and the shape and size of the chamber. The excavation depth: Usually, the greater the buried depth, the higher the geostress. As has been mentioned before, the high geostress is the main external factor affecting the occurrence of rockburst, considering the excavation depth is also an important aspect in studying the influencing factors of rockburst, since the excavation depth is also closely related to the change of geostress. The shape and size of the chamber: By analyzing the stress states in the circular chamber and non-circular chamber (straight wall round arch), we find, under the same stress environment, the stress concentration effect in the circular chamber is less than that in the non-circular chamber. Therefore, the intensity and probability of rockburst in the circular chamber are lower than those in the non-circular chamber. Thus it can be seen that the shape and size of the excavation chamber are closely related to the occurrence of rockburst.

②The influence of hydrology and engineering geology on rockburst. The statistical results show that the occurrence of rockburst is related to the tectonic activities. In the area near the fold axis or in the fault zone, where the geostress is concentrated or there exists huge residual tectonic stress, the rockbursts occur densely. When the work face is close to the fault or the fold axis, the frequency and intensity of rockbursts increase. Moreover, the rockbursts generally occur in the area where groundwater is not developed.

③Blasting, earthquake and other inducing factors. Blasting or earthquake is an important external cause of rockburst. In the event of a blast or an earthquake, the produced huge elastic wave generates a dynamic response to the surrounding cave, which exacerbates the stress release in the surrounding rocks, causing the rock mass in the critical state, to be disturbed, resulting in sudden instability or damage. So the rockburst there would be more serious. According to some incomplete statistics, about 20% of rockbursts are caused by blasting or earthquake or other related factors.

In general, rockbursts are caused by excavation of chambers or tunnels, where the in-situ geostress differentiates, with stress in the surrounding rock jumping-up, and the energy concentrated. Under the action of stress in surrounding rock, some tensile-shear brittle failure occurs, accompanied by sound and vibration. The surrounding rocks convert from static equilibrium to dynamic instability, which causes the rock pieces (blocks) to isolate from the parent body, to obtain effective ejection energy, and to throw (bounce, scatter) violently into the air. The suddenness and uncertainty of rockbursts have brought huge obstacles to the TBM construction. Through the analysis of the influencing factors of rockburst, we hope to provide some guidance to the subsequent survey and construction.

7.3 Research Status of Advanced Rockburst Forecast Technology

Since the record of rock burst occurrence in the Leipzig coal mine in Britain in 1738, there have been reports of rock bursts in various countries around the world. In recent years, based on the research of rockbursts, Tang Jiang, Huohuo Li, and Handong Liu summarized the current rockburst theories, analyzed the advantages and disadvantages of various rockburst theories, and briefly forecasted the developing trend of rockburst research. Shaohui Tang and Linsheng Xu introduced the rockburst phenomenon in the actual engineering case and the results of field research. Yuanhan Wang and others proposed some methods to judge or discriminate the possibility and scale of rockbursts. Tianbin Li and others introduced some research about the physical simulation tests of rockburst in China. After synthesizing these literature, we find:

①Scholars at home and abroad have studied rockbursts from various perspectives, but the basic understanding of rockbursts is still different. To sum up, there are basicly two views of the definition of rockburst currently: one, represented by the Norwegian expert B.F.Russense, holds that if there are sound, slope, burst, spalling and even ejection of rocks, with fresh fracture surfaces, this can be called a rockburst; the other, represented by Yian Tan, a Chinese scholar, holds that only by producing throwing failure can it be called a rockburst, while those phenomena of non-dynamic ejection and indoor deformation and fracture are classified as static brittle failure. Professor Xieyuan Zhang and others divided rockbursts into three types, according to the location and the amount of energy released: the rockburst caused by sudden rupture of the surface rock of surrounding cave, the rockburst caused by sudden failure of pillar or large-scale surrounding rock, and the rockburst caused by fault dislocation. According to the cause of high geostress in the burst rock mass, Linsheng Xu and Lansheng Wang firstly divide rockbursts into four types: the gravity stress type, the tectonic stress type, the variation stress type and the comprehensive stress type; then according to the specific stress conditions and the characteristics of rockbursts, they further divide them into 8 subtypes.

②At present, there are many theories used for analyzing rockbursts: the strength theory, the energy theory, the stiffness theory, the fracture damage theory, the catastrophe theory and so on. In addition, some Chinese researchers have studied the localization of rock deformation and the stability of rock mechanics system, by using the bifurcation theory, the dissipative structure theory and the chaos theory, which have greatly promoted the development of the rockburst theory and the rock instability theory in China. The understanding of mechanism of rockburst is mainly from the perspectives of stress and energy. From the perspective of stress, some researchers think that rockburst is caused by excavation and unloading, and others hold that rockburst is caused by stress concentration, i.e. the increasing load; from the aspect of energy, rockburst is considered to be caused by the sudden release of elastic strain energy, stored in rock mass after excavation.

③Traditionally, the forecast methods of rockbursts mainly include: prediction of comprehensive geological analysis methods, σ_0/R_b criterion forecast method, the rockburst storage test analysis and forecast method, rockburst critical depth forecast method, acoustic emission site monitoring forecast method, electromagnetic radiation monitoring and forecasting method. In recent years, Chun'an Tang and other researchers have developed the RFPA numerical calculation tools for rock failure process analysis with China's independent copyright, devised an ESG microseismic monitoring system, and configured a visualizotion software RFPA. View. View software, which provide new research methods for the study of the instability of rock failure, including rockbursts. Yingchun Yang and Jing Zhu proposed an extendable evaluation method. Tong Jiang and others put forward the application of the optimal classification model in grey system. Xiating Feng and Hongbo Zhao proposed a support vector machine method for rockburst classification. All these methods provide references to the forecast of rockburst.

7.4 De-stress Blasting Test Under Double Shield TBM

The test site is selected at a highway tunnel in Tibet, with double shield TBM construction. The project is located in the southeast of the Qinghai-Tibet Plateau and is the most uplifted and eroded area in the Qinghai-Tibet Plateau, with undulating terrain and deep valley. The mountains there above 4,000 m are mostly covered with snow and ice; down below 3,200 m they are mostly covered by virgin forests. The lone block of moraine is the main road, and the local bedrock steep ridges are exposed intermittently. The lithology is dominated by mixed gneiss, with strong physical and geological effects such as freezing and thawing, weathering and unloading of bank slopes.

The rock in the tunnel area is dominated by mixed gneiss, and the rock belongs generally to medium to hard rock, with few extremely hard rocks. The anisotropy of rock strength affected by gneiss is obvious. The bedrocks at the entrance and exit are exposed, and the underlying gneisses of the Duoxuanlayan Formation are mainly weakly weathered. The weathering of rock mass is strong, the unloading depth of the rock mass is 30–40 m, and the weak unloading depth is 70–80 m.

The diameter of excavated tunnel is 9.13 m, the lining adopts "6 + 1" type quadrilateral prefabricated concrete segments (Figure 7.14 and Figure 7.15), with a thickness of 35 cm, and the diameter of tunnel after lining is 8.1 m, the tunnel is a two-lane tunnel, with a construction boundary width of 7.0 m to 4.5 m.

Figure 7.14 Departure photos of double shields

Figure 7.15 "6 + 1" quadrilateral precast concrete segment

The entrance elevation of the tunnel is 3,547 m, the exit elevation is 3,566 m, and the maximum buried depth is about 832 m. The surrounding rock in the tunnel is mainly mixed gneiss and granite gneiss, the average compressive strength is 75–90 MPa, the quartz content is 15%–30%, and the wear resistance index is between 4–5.5. Due to the strong new tectonic movement in the tunnel area, the structural in-situ stress is of high magnitude under the structural compression. At present, the measured value of the maximum principal stress is 27.2 MPa.

The tunnel of this project is currently driven by double-shield TBM. Double-shield TBM construction mainly uses segment lining. After lining, the bean gravel is backfilled between the segment and surrounding rock and then grouted to make the bean gravel grouting as a whole.

The Control Method on the Pre-release De-stress Blasting for High-intensity Rockburst

During the construction of the tunnel, there were many strong rockbursts, which caused severe jamming 5 times, all of which were located in the interlayer zone in the obvious weak surface. Therefore, based on the geological data, a transition zone was selected as the test tunnel section. The experimental process is as follows.

7.4.1 Native stress test

Stop the cutter head, use TBM's advanced drilling rig to drill forward. The processes are the following: ①Drilling in-situ stress test holes. To drill a large hole first, the parameters are: the drilling diameter 130 mm, the elevation angle 15°, the depth 10 m; then drilling the conical hole: after reaching the design depth, observe the core taken, if there is no crack, we can start drilling the conical hole. Use a flat drill to smooth the bottom of the hole, and use a conical drill to drill the bell mouth. In order to facilitate the next step of drilling concentric small holes, clean drilling holes and probes to enter the small holes smoothly; then drilling small holes: drilling a concentric hole with a diameter of 36 mm from the bottom of the hole, the depth of the small hole is 32–40 cm. After punching the holes, rinse them with water and send the wiping head soaked in acetone into the holes to scrub back and forth to completely remove the oil and other dirt in the holes and dry them. ②Installation of hollow core package. Sand the outer cylindrical surface of the hollow core package with sandpaper; prepare the adhesive (two liquid materials A and B) in proportion, pour the appropriate amount of adhesive into the cavity of the hollow core package, and fix the pins. Install the package on the director. Use a drill pipe to slowly feed it into the large hole, and note down the length. Special care should be taken when the remaining length is about 5 m, to ensure that the package can enter the small hole completely. The front end is about 20 cm into the small hole. It should be noted that the barrel part of the package body is slowly pushed in, the probe and the installer are uncoupled, and the package body is successfully installed in the small hole. ③Initial strain data. About 20 hours after the package is installed, the epoxy resin is cured. Carefully lift the installer out of the drill hole. The electronic compass in the installer writes down the installation angle of the probe. The number shown is the installation angle of the stress gauge. ④Stress release. Use the first step of the thin-wall drill for large holes to continue to deepen the large holes, so that the core around the small holes can achieve stress release. The deformation or strain of the small hole caused by the stress release is measured by the measurement system, including the test probe, and recorded by the recording instrument. According to the measured deformation or strain of the small hole, the state of in-situ stress around the small hole can be obtained through the relevant formula. ⑤ On-site rock confining pressure test. The confining pressure is set to 5 levels of 0, 2, 4, 8 and 12 MPa, and the strain value under each level of confining pressure is obtained. ⑥Elastic modulus test and data processing.

7.4.2 De-stress bursting test

After calculation of the initial stress, use a rock drill, equipped with the double shield TBM equipment, to drill a total of 7 blast holes, (the total number of holes can be up to 14 holes). The blasting test parameters are shown in Table 7.1.

Table 7.1 Blasting parameter table

Number of holes	7
Borehole depth	10 m
Hole diameter	50 mm
Uncoupling coefficient	1.0
Initiation method	Positive detonation
Single hole charge	2.0 kg
Borehole spacing	1 m

The layout and side view of de-stress blasting holes are shown in Figure 7.3 and Figure 7.4. The location of the TBM ground stress blasting hole in the tunnel is shown schematically in Figure 7.16.

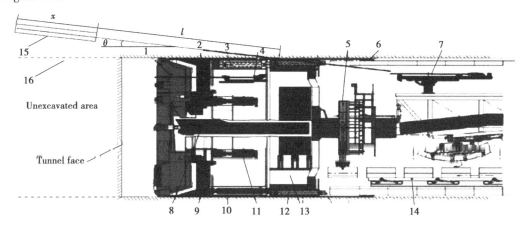

Figure 7.16 Schematic diagram of TBM de-stress blasting hole location

1—Cutter head 2—Main bearing 3—Main propulsion cylinder 4—Multi-function drilling rig
5—Tube piece installation machine 6—Shield tail seal 7—Advanced drilling rig 8—Belt conveyor
9—Front shield 10—Telescopic cylinder 11—Cutter head drive 12—Auxiliary propulsion cylinder
13—Support shield 14—Segment conveyor 15—Charge explosives 16—Excavation extension surface
x—Charge explosive length l—Length of unfilled explosives

After the drilling is completed, shrink the cylinder and move the cutter head back 0.5–1 m, even if there is a gap of 0.5-1 m between the cutter head and the palm face, then perform the following steps: Hole inspection, the hole inspector, assigned by the team leader shall, follow the blasting design. Acceptance by holes, some remedial measures should be taken in time for unqualified holes, according to the opinions of engineering and technical personnel; Charge, the operation should be carried out in accordance with the technical design in process of charging, and use gun sticks or ropes while charging. Measure the height of the drug to avoid plugging in the lower part; After filling, the slag around the drill hole is used to fill the blast hole. Stones and flammable materials should not be used for plugging; Detonation, detonation work should be done by the experienced blaster, designated by the team leader, to conduct a detailed inspection of the blaster and other equipment before initiation, to ensure the smooth implementation of the initiation. After the inspection, the inspection work is carried out by the blasting captain and the experienced blaster, and the inspection will be carried out after the blasting dust is dispersed. The inspection area is roughly the same as the schematic diagram of de-stress blasting and crushing zone, in Figure 7.17.

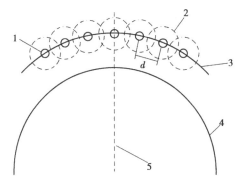

Figure 7.17　Schematic diagram of de-stress blasting and crushing zone

1—Explosives　2—Blasting affected area　3—Circumferential axis of explosives
4—Surrounding rock wall　5—Tunnel axis　d—Blasthole spacing

After the blasting work, stress test again according to step 7.4.1 and record the test data.

7.4.3　Data processing results

Before blasting, the depth of the test point is 10 m above the slope in the front of the palm. After data processing and calculation, the calculation results of the in-situ stress test are shown in Table 7.2.

Table 7.2　In-situ stress test results before blasting

Test site	Lithology	σ_1/MPa	σ_2/MPa	σ_3/MPa
K1+386	Gneiss	21.8	16.4	13.2

After blasting, the test point is set at the relatively complete surrounding rock mass on both sides of the broken ring, and the depth is 10 m above the slope in the front of the tunnel face. After data processing and calculation, the results of in-situ stress test calculation are shown in Table 7.3.

Table 7.3 In-situ stress test results table after blasting

Test site	Lithology	σ_1/MPa	σ_2/MPa	σ_3/MPa
K1+386	Gneiss	10.1	8.7	8.1

The compressive strength σ_c of the stress test point is 109.3 MPa, and the pre-blast discrimination coefficient is $\sigma_c/\sigma_1 = 5.01$, which belongs to the moderate rockburst grade. After de-stress blasting, the pre-blast discrimination coefficient is $\sigma_c/\sigma_1 = 10.82$, which belongs to the low rock burst level. The data processing results show that the maximum principal stress before blasting is 21.8 MPa, the maximum principal stress after blasting is 10.1 MPa, and the maximum stress decreases by 54%. Thus blasting has a significant effect on the reduction of the maximum principal stress. According to the aforementioned rockburst classification standards, it belongs to the moderate rockburst before blasting. After decompression blasting, the rockburst reaches a low level or even close to no rockburst activity. The stress-release effect of de-stress blasting is remarkable.

7.5 Summary

①Taking the double shield TBM construction in the tunnel of a highway in Tibet as the test site, we summarize the rock burst problems, encountered in the double shield TBM construction, and analyze the geological information collection and the advanced geological forecast under the double shield TBM construction. After discussing the influencing factors of various means, and evaluating the suitability of the existing forecasting methods in the TBM construction process, we hope to improve the existing geological forecasting methods, such as the geophysical prospecting and advanced drilling, and to propose a multi-scale and multi-way dynamic advanced geological forecasting method, so as to greatly promote the technical progress of the long and deep buried tunnels.

②Our research also aims to improve the geological advanced forecast technique for the double shield TBM construction in tunnel. For the first time, a geophysical test method of using a new source excitation, and a rapid and efficient method for installing geophone was proposed, and the interpretation criteria for geophysical test, based on numerical simulation, were established, and the advanced geological drilling technique under continuous tunneling conditions was realized, and a comprehensive forecast method was formed by means of fuzzy math. A technological device for advanced geological drilling, under continuous tunneling conditions, was developed. In order to solve the problems of low drilling efficiency and low speed of drilling rig, we proposed the following ideas: a downtime period of equipment maintenance and repair, the advanced drilling depth, the number of holes required for geological forecasting, a scheme for optimizing and improving the structure of the cutter head. Cutter parts that can be synchronized with holes are used to achieve synchronized holes during TBM construction. In addition, to reduce the construction process without causing interference to the work of various professionals, the thrust of the tunneling machine itself is used to make holes. The quality is also guaranteed, and the hole making tasks with different apertures and lengths can be completed, according to actual needs.

③The de-stress blasting technology was applied to the TBM tunnel, and a verification test was carried out. The experimental results showed that the maximum stress decreased by 54%, thus the blasting had a significant effect on the reduction of the maximum principal stress. According to the aforementioned rockburst classification standards, it belongs to moderate rockburst before blasting. After de-stress blasting, the rockburst reaches a low level or even close to no rockburst activity. The stress-release effect of de-stress blasting is remarkable.

Conclusion and Outlook

In these years, with the implementation of many major projects in the mountainous areas of western China, rock bursts, encountered by deep underground engineering, have become more frequent and led to many casualties, equipment losses and project delays, bringing serious obstacles to the projects. For those high-intensity rock burst hazards, though many scientific research have been published, there are no effective measures to control them at present. The existing methods, such as the high-pressure water injection method, the benching method, the reinforcement treatment method, and the borehole unloading method, etc., only have certain effects on the rockbursts of moderate or minor degree, while their effect on high-intensity rock bursts is not obvious. Thus, focusing on the high-intensity high-risk rock bursts and based on previous studies, authors of this book analyze the mechanism of the high-intensity rock bursts, and propose a "blasting in response to blast" method, that is, to drill advanced blastholes to implement explosion before the excavation of the tunnel face, which, by forming a loose broken ring outside the excavation line, can redistribute or release the stress in the tunnel walls to achieve early geostress release. Since the selection and optimization of blasting parameters have an impact on the de-stress effect and the stability of the tunnel after excavation, this book, based on the mechanism of advanced de-stress blasting on the high-intensity rock bursts, applies the numerical simulation methods to select reasonable blasting parameters, including the diameter of the blasthole, the distance between the blastholes, etc, and the on-site test method to verify the validity of the blasting method and the rationality of the blasting parameters.

The main conclusions are can be drawn as follows:

Through the numerical simulations of the mechanical properties of the surrounding rocks and the tunnel excavation conditions, comparing various rockbursts control measures, a conclusion can be drawn that the compressive strength, the angle of internal friction, the elastic modulus, and the cohesive forces are the main influencing factors of rock burst activity. Through numerical simulation of the excavation conditions, we find: with the increasing sizes of excavation, the unloading space for excavation is also increasing, that is, there are more spaces for the redistribution of stress in the surrounding rocks in tunnel, which inevitably leads to a decrease in σ_θ, thus resulting in a decrease in σ_θ/σ_c. Therefore, with the increasing sizes of excavation, the rock burst activities will gradually weaken. As to the step-by-step excavation, the σ_θ/σ_c after the upper excavation is reduced, compared to the full-section excavation, while after the lower excavation, σ_θ/σ_c is larger than the upper excavation but still smaller than the σ_θ/σ_c of the full-section excavation, which proves that the simple lining support, to connect the surrounding rocks as a unified whole to jointly bear the geostress, has a certain inhibitory effect on the rock burst activity. For the control of the high-intensity rock bursts, the current advanced de-stress prerelease blasting method is applicable, since this method takes advantage of the huge energy from blasting vibration or shock waves to disturb the surrounding rocks, guiding the re-distribution of stress and partial releasing the stress, thereby reducing the intensity of rock bursts.

After deriving the residual stress calculation formula of the de-stress blasting, the stress value at the corresponding point in the tunnel wall after de-stress blasting can be calculated. The residual stress equations for de-stress blasting under three-dimensional conditions are:

$$\begin{cases} \sigma_{Vr} = \sigma_v - k\rho_r C_p Q^\alpha (r_1^{-\alpha} \cos \delta_1 + r_2^{-\alpha} \cos \delta_2 + \cdots + r_n^{-\alpha} \cos \delta_n) \\ \sigma_{Zr} = \sigma_z - k\rho_r C_p Q^\alpha (r_1^{-\alpha} \cos \beta_1 + r_2^{-\alpha} \cos \beta_2 + \cdots + r_n^{-\alpha} \cos \beta_n) \\ \sigma_{Hr} = \sigma_H - k\rho_r C_p Q^\alpha (r_1^{-\alpha} \cos \gamma_1 + r_2^{-\alpha} \cos \gamma_2 + \cdots + r_n^{-\alpha} \cos \gamma_n) \end{cases} \quad (6.14)$$

Where σ_{Vr} is X axial residual stress value at the de-stress point, σ_{Zr} is Z axial residual stress value at the de-stress point, σ_{Hr} is Y axial residual stress value at the de-stress point, α is the stress decay index is generally 1 to 2, C_p is the elastic wave velocity in the rock, σ_v is X axial stress value, σ_z is Z axial stress value, σ_H is Y axial stress value, $r_1, r_2, \cdots, r_n$ are the distance from the center of each drug room to the de-stress point, where

$$r_n = \sqrt{(x - x_n)^2 + (y - y_n)^2 + (z - z_n)^2}$$

$\delta_1, \delta_2, \cdots, \delta_n, \beta_1, \beta_2, \cdots, \beta_n, \gamma_1, \gamma_2, \cdots, \gamma_n$ are the angle between the blasting stress of each blasthole and the x-axis, y-axis and z-axis at the de-stress point, wherein

$$\begin{cases} \cos \delta_n = \dfrac{|x - x_n|}{\sqrt{(x - x_n)^2 + (y - y_n)^2 + (z - z_n)^2}} \\ \cos \beta_n = \dfrac{|y - y_n|}{\sqrt{(x - x_n)^2 + (y - y_n)^2 + (z - z_n)^2}} \\ \cos \gamma_n = \dfrac{|z - z_n|}{\sqrt{(x - x_n)^2 + (y - y_n)^2 + (z - z_n)^2}} \end{cases}$$

ρ_r is the density of the rock, K is the coefficient associated with rock properties, blasting process parameters, and Q is the amount of charge.

The de-stress blasting of surrounding rocks under high geostress conditions was simulated by numerical simulation software ANSYS and FLAC. According to the actual situation, the numerical simulation parameters of the de-stress blasting were analyzed and determined, mainly including 7 parameters and value intervals, such as the diameter of the blasthole, the depth of the blasthole, the coupling coefficient of the blasthole, the detonation sequence, the angles between the blasthole and the axis of the tunnel on the X direction and on the Y direction. Firstly, the orthogonal experiment method was used to design and determine the blasting scheme of the de-stress blasting under multi-parameter interaction conditions. Then the blasting scheme of the de-stress blasting under changing conditions of single factor was designed, and the results were analyzed. Based on the numerical simulation analysis of each experiment in the orthogonal simulation method, and combined with the evaluation of discriminant coefficients of the de-stress blasting, we first determine the appropriate value range of each parameter, under the influence of mutual changes: the length of the blasthole is 5 – 8 m, the distance between the blastholes is 20 – 70 cm, the uncoupling coefficient is 1.0 – 2.5, the diameter of the blasthole is 50 – 70 mm, and the angle between the blasthole and the tunnel in the X direction is $15° - 50°$, the angle between the blasthole and the tunnel in the Y direction is $15° - 60°$, and the initiation mode is the reverse initiation. Then according to the actual blasting situation and the on-site numerical simulations, a set of optimal combination of parameters is determined: the length of the blasthole is 5 m, the distances between blastholes is 70 mm, the non-coupling coefficient is 1.5, the blasthole diameter is 50 mm, and the initiation mode is the reverse initiation.

Based on the theoretical analysis and the numerical simulation results, an on-site de-stress blasting test was carried out at K186+94.020 of the Sangzhuling tunnel of Linla Railway. The optimal combination of parameters for numerical simulation was applied, that is, the length of the blasthole is 5 m, the distance between blastholes is 70 mm, the non-coupling coefficient is 1.5, the blasthole diameter is 50 mm, and the detonation mode is the reverse detonation. The test results show that the maximum principal stress before blasting is 30.7 MPa, while the maximum principal stress after blasting is 11.5 MPa, and the maximum stress is reduced by 63%, which indicates that blasting does have a significant effect on the reduction of the maximum principal stress.

The measured data are compared with the formula calculation results to verify the correctness and reliability of the formula. Through the residual stress formulas derived in this book, after the de-stress blasting, we find the calculation result is close to the measured maximum stress value, which is slightly various but acceptable, and which does not affect the discrimination and classification of the rock burst intensity. The ratio of residual stress to compressive strength is less than 0.3, which is a non-rock burst activity, indicating that the formula calculation result is credible and consistent with the on-site test. Comparing the calculation results of the formula with results of on-site numerical simulation, the figures are consistent, indicating that the derived formulas are reliable, and the above-mentioned methods are mutually verified.

The de-stress blasting technology was applied to the TBM tunnel, and a verification test was carried out. The experimental results showed that the maximum stress decreased by 54%, thus the blasting had a significant effect on the reduction of the maximum principal stress. According to the aforementioned rockburst classification standards, it belongs to a moderate rockburst before blasting. After the de-stress blasting, the rockburst reaches a low level or even close to no rockburst activity. The stress-release effect of the de-stress blasting is remarkable.

According to the aforementioned rock burst grading standard, it is a moderate rock burst before blasting, and after de-stress blasting, it reaches a low level; or even close to no rockburst activity. The release effect of the de-stress blasting was obvious and reached the expectations, which indicates that the combination of parameters is scientific and reasonable. Thus this method is an active pre-release measure, featured by advanced stress release prior to tunnel excavation, and can be used in the prevention of high-intensity rock bursts.

This book relies on the natural science fund project "Study on the method of in-situ stress pre-release controlled blasting for high-strength rock blasting" (No. 41572358). In finishing the project, the authors really understood the meaning of the phrase "rock burst governance is a worldwide problem". What the authors have done just touched the epidermis of the field. The authors hope that this book will inspire more colleagues' to pay attention to the disaster-causing mechanism of rockbursts in engineering projects. Personally, there is still much work to be done in the following aspects:

The influencing rules of blasting parameters and the optimization of parameter combination are based on the numerical simulation, which cannot be wholly verified and optimized through on-site tests, since it would require hundreds of blasting tests and stress tests, which require more financial support and may delay the construction period of the tunnel under construction.

Because the technological parameters of drilling and blasting are in combination, and the variables interact with each other, the quantification of the influencing factors is difficult to express directly in formulas with limited parameters. The accuracy of the process influence coefficient k is insufficient. The formulas also cannot explain the phenomenon that the principal stress direction is deflected after de-stress blasting, so further optimization is still subject to more fitting and correction.

The timeliness of the rock burst will also affect the choice of de-stress measures and the de-stress effect. Though there are some records of the on-site rockburst frequency, the influencing mechanism of de-stress has not been studied in depth.

At present, there are a large number of tunnels and underground projects on construction in high-stress areas in China, which provide some precious researching sites and cases for deep engineering research. The above-mentioned issues to be further studied will prompt the prevention of rock bursts through the de-stress blasting method. The authors hope the peers will put more effort and gradually overcome those various frontier problems in the field of rock burst prevention.

Acknowledgements

In the process of writing this book, we have received strong support from many units and selfless colleagues and friends. Here, we would like to express our sincere gratitude and deep respect for them. The School of Environment and Civil Engineering of Chengdu University of Technology has a national first-class professional teaching team, which has cultivated many excellent talents, through which we are able to work with many famous teachers and receive their help. Our colleagues in the tunnel and underground engineering team of State Key Laboratory of Geohazard Prevention and Geoenvironment Protection have provided us with a lot of help, especially the team leader, Professor Tianbin Li, who did not hesitate to write, and personally worked many time to complete it. Professor Tianbin Li's rigorousness in study, forgiving mind, and humility have deeply influenced us and become a valuable asset in our life.

We would like to thank Chief Engineer Zili Pan and Chief Engineer Qingsong Liu of China Railway Second Academy, Chief Engineer Junfeng Yang of Lin-la Railway Project Headquarters of China Railway No. 5 Bureau, and the colleagues at Lin-la Project Department of China Railway No.5 Bureau for their selfless help. They provide on-site test sites and various facilities to allow our tests to be successfully completed in the harsh conditions of the plateau.

In the process of completing this article, our colleagues Xiangli Guan, Dr. Zhenlin Chen, Dr. Qian Li, and Dr. Mingming Zheng also gave us strong support. Thank you very much. We also want to thank our graduate students Wang Peng, Su Tao, Zhao Hua, Hu Gaoyuan, Ming Junnan, Zeng Yifang, Zhang Ningxin, Song Yu and other classmates. They actively participated in the process of experiments, data processing and thesis writing, and paid a lot of labor.

This book mainly records some of the author's experiences in the treatment of rockburst disasters in deep and long tunnels buried in Tibet in recent years for the reference of colleagues. Due to the limited time, some errors are unavoidable, ineuitable please criticize and correct us. May this book not waste reader's precious time!

References

ADOKO A C, GOKCEOGLU C, WU L, et al., 2013. Knowledge-based and data-driven fuzzy modeling for rockburst prediction [J]. International Journal of Rock Mechanics & Mining Sciences, 61(4): 86-95.

BOLER F M, SWANSON P L, 1993. Seismicity and stress changes subsequent to de-stress blasting at the Galena Mine and implications for stress control strategies [M]. Washington, D. C.: Department of the Interior, Bureau of Mines.

CAI J D, LIU J H, LI H M, 2008. Application of Blasting Pressure Relief Technology in Rock Burst Prevention and Control[J]. Blasting, 25(1): 1-4.

CHEN G Q, LI T B, HE Y H, et al., 2013. Unloading thermal-mechanical effect and rockburst trend analysis of deep-buried hard rock tunnels[J]. Chinese Journal of Rock Mechanics and Engineering, 32(8): 1554-1563.

CHEN Z W, 2006. Research on the occurrence mechanism and preventive measures of rock burst in Dataijing rock tunnel[D]. Fuxin: Liaoning Technical University.

Chen Z J, 1987. Engineering records, theory and control of rock burst [J]. Chinese Journal of Rock Mechanics and Engineering, (1): 9-26.

DONG Z X, SHAO P, 2005. Blasting Engineering[M]. Beijing: China Construction Industry Press.

DU Z J, 2007. Rockburst prediction theory and application research [D]. Wuhan: Wuhan University.

FEIT G N, MALINNIKOVA O N, ZYKOV V S, et al., 2002. Prediction of rockburst and sudden outburst hazard on the basis of estimate of rock-mass energy [J]. Journal of Mining Science, 38(1): 61-63.

FENG S D, 2007. The application of deep hole pressure relief blasting in dynamic pressure roadway[J]. Shanxi Architecture, 33(19): 118-119.

FENG T, 1999. Rockburst mechanism and prevention theory and application research [D]. Changsha: Central South University.

FENG X T, 2013. Dynamic design method of deep-buried hard rock tunnel[M]. Beijing: Science Press.

GAY N C, JAGER A J, RYDER J A, et al., 1995. Rock-engineering strategies to meet the safety and production needs of the South African mining industry in the 21st century [J]. Journal of the Southern African Institute of Mining and Metallurgy, (3): 115-136.

GERMANOVICH L N, DYSKIN A V, 2000. Fracture mechanisms and instability of openings in compression [J]. International Journal of Rock Mechanics & Mining Sciences, 37(1): 263-284.

GONG F Q, LI X B, 2007. The distance discrimination method and application of rockburst occurrence and intensity classification prediction[J]. Chinese Journal of Rock Mechanics and Engineering, 26(5): 1012-1018.

GU J C, FAN J Q, KONG F L, et al., 2014. Throwing Rockburst Mechanism and Simulation Test Technology[J]. Chinese Journal of Rock Mechanics and Engineering, 33(6): 1081-1089.

GUO L, WU A X, HU J G, et al., 2003. Rock mechanics problems and analysis methods of deep hard rock mining[J]. Metal Mine, (3): 4-7.

GUO X C, 2010. Research on pressure relief and support technology for deep buried high ground stress tunnels[D]. Xi'an: Xi'an University of Science and Technology.

GUO Z, 1996. Practical Rock Mass Mechanics[M]. Beijing: Seismological Press.

HAJIABDOLMAJID V, KAISER P K, 2003. Brittleness of rock and stability assessment in hard rock tunneling[J]. Tunnelling and Underground Space Technology, 18(1): 35-48.

HAJIABDOLMAJID V, KAISER P K, MARTIN C D, 2002. Modelling brittle failure of rock [J]. International Journal of Rock Mechanics and Mining Sciences, 39(6): 731-741.

HONG K R, 2015. Development status and prospects of tunnels and underground engineering in my country[J]. Tunnel Construction, (2): 95-107.

HU W Q, ZHENG Y R, ZHONG C Y, 2004. Construction method and numerical analysis of the weak surrounding rock section of Muzhailing Tunnel[J]. Chinese Journal of Underground Space and Engineering, 24(2): 194-197.

HUANG R Q, WANG X N, 1998. Analysis of the main hazard geological problems of deep-buried tunnel engineering[J]. Hydrogeology and Engineering Geology, (4): 21-24.

INDUSTRY STANDARDS COMPILATION GROUP OF THE PEOPLE'S REPUBLIC OF CHINA, 1999. Technical Specifications for Underground Excavation Engineering of Hydraulic Buildings: DLT5099-1999 [S]. Beijing: Hydraulic and Electric Power Press.

JIANG T, LI H Y, LIU H D, 1998. Research status of rockburst theory[J]. Journal of North China Institute of Water Conservancy and Hydropower, (1): 46-48.

JIANG Y, 2002. High in-situ stress rock burst and karst water inrush problems and countermeasures in deep-buried long highway tunnels[J]. Chinese Journal of Rock Mechanics and Engineering, 21(9): 1319-1323.

JING H W, LI Y H, XU G A, 2005. Research on stability analysis and control technology of surrounding rock of deep buried roadway[J]. Rock and Soil Mechanics, 26(6): 877-880.

JING H W, FU G B, 1999. Field measurement analysis and control technology research on influencing factors of surrounding rock loose zone in deep mine roadway[J]. Chinese Journal of Rock Mechanics and Engineering, 18(1): 70-74.

KEERTHI S S, SHEVADE S K, BHATTACHARYYA C, et al., 2001. Improvements to Platt's SMO algorithm for SVM classifier design [J]. Neural computation, 13(3): 637-649.

LINKOV A M, 1996. Rockbursts and the instability of rock masses [J]. International Journal of Rock Mechanics and Mining Sciences & Geomechanics Abstracts, 33(7): 727-732.

MAZAIRA A, KONICEK P, 2015. Intense rockburst impacts in deep underground construction and their prevention [J]. Canadian Geotechnical Journal, 52(10): 1426-1439.

ORTLEPP W D, 1994. Grouted rock-studs as rockburst support: a simple design approach and an effective test procedure [J]. Journal of the Southern African Institute of Mining and Metallurgy, (2): 47-63.

ORTLEPP W D, 1997. Rock fracture and rockbursts: an illustrative study [M]. Johannesburg, South Africa: South African Institute of Mining and Metallurgy Press.

ORTLEPP W D, STACEY T R, 1994. Rockburst mechanisms in tunnels and shafts [J]. Tunnelling & Underground Space Technology, 9(1): 59-65.

RUSSENES B F, 1974. Analysis of rock spalling for tunnels in steep valley sides [D]. Trondheim: Norwegian Institute of Technology.

SILENY J, 1986. Inversion of first-motion amplitudes recorded by local seismic network for rockburst mechanism study in the Kladno, Czechoslovakia, mining area [J]. Acta Geophysica Polonica, 34(3): 201-213.

SINGH P K, ROY M P, PASWAN R K, 2014. Controlled blasting for long term stability of pit-walls [J]. International Journal of Rock Mechanics and Mining Sciences, 70: 388-399.

TANG B, 2000. Rockburst control using de-stress blasting [D]. Montreal: McGill University.

YAN P, ZHAO Z, LU W, et al., 2015. Mitigation of rock burst events by blasting techniques during deep-tunnel excavation [J]. Engineering Geology, 188: 126-136.

HOEK E, BROWN E T, 1986. Rock Underground Engineering [M]. Beijing: Metallurgical Industry Press, 5-15.

LI D H, WANG D P, GAO B B, 2005. Finite element simulation analysis of viscoelastic model of surrounding rock[J]. Mining and Metallurgical Engineering, 25(1): 1-2.

LI J F, HE G, GUO X F, 2008. The mechanism and application of deep hole pressure relief blasting to control rock burst[J]. Coal Science and Technology, (1): 75-77.

LI T B, WANG X F, MENG L B, 2011. Physical simulation of similar materials for rockburst [J]. Chinese Journal of Rock Mechanics and Engineering, (S1): 2610-2616.

LI Y F, 2015. Research on Risk Assessment and Countermeasures of Research Report on TBM Construction Tunnels in Plateau[D]. Chengdu: Chengdu University of Technology.

LI Y S, 1982. Research and review of mine shock at home and abroad[J]. Reference Materials for

Coal Research, (4): 1-10.

LIU G, SONG H W, 2003. Numerical simulation of influencing factors of surrounding rock loose zone[J]. Mining and Metallurgical Engineering, 23(1): 1-3.

LIU M S, YANG J, GE W H, et al., 2010. Research on engineering treatment methods for rockburst disasters in the diversion tunnel of Jinping II Hydropower Station [J]. Guizhou Hydropower, 24(5): 11-15.

LU J Y, 1993. Research status of several issues in rockburst prediction and prevention[J]. Design of Hydropower Station, 9(1): 55-59.

LUO Y, SHEN Z W, 2006. Deep hole controlled pressure relief blasting mechanism and outburst prevention test research[J]. Chinese Quarterly of Mechanics, 27(3): 469-475.

MA W W, LI J T, LIANG W X, et al., 2015. Numerical analysis of rock failure under explosive stress wave[J]. Coal Mine Safety, 46(9): 188-191.

PAN T L, 1996. Experimental study on loosening blasting pressure relief of roadway surrounding rock[J]. Northeast Coal Technology, (4): 22-25.

PAN Y S, ZHANG M T, LI G Z, 1994. Sharp-angle catastrophe model of cavern rockburst [J]. Applied Mathematics and Mechanics, (10): 893-900.

QIAN J H, YIN Z Z, 1996. Geotechnical Principles and Calculations (Second Edition) [M]. Beijing: China Water Power Press.

QIN Y, 2013. Research on stability of surrounding rock and bolt parameters of tunnel with high risk of rockburst at large depth[D]. Chengdu: Southwest Jiaotong University.

SU B, 2015. Research on parameters of pressure relief controlled blasting in high ground stress tunnels[D]. Chengdu: Chengdu University.

SUN R Z, FANG G D, 2015. Rockburst prediction and prevention in tunnel construction [J]. Civil Engineering Technology and Design, (10): 948.

TAN Y A, 1992. Classification of Rockburst Intensity[J]. Geological Review, 38(5): 439-443.

TANG S H, 2006. Research status of rockburst disasters in deep metal mines[J]. Mining Research and Development, 26(S1): 136-140.

TANG S H, WU Z J, CHEN X H, 2003. Research on the Occurrence and Formation Mechanism of Rock Burst in Deep Underground Mines [J]. Chinese Journal of Rock Mechanics and Engineering, 22(8): 1250-1254.

WANG B, HE C, YU T, 2007. Numerical analysis of Cangling tunnel rockburst prediction and discussion on the timing of initial support[J]. Rock and Soil Mechanics, 28(6): 1181-1186.

WANG Y, WANG J M, YIN J M, et al., 2012. Research on rockburst prevention measures for deep-buried tunnels based on rapid stress release[J]. Rock and Soil Mechanics, 33(2): 547-553.

WEI M Y, WANG E Y, LIU X F, et al., 2011. Numerical simulation study on the effect of pressure relief blasting in deep coal seam to prevent rock burst[J]. Rock and Soil Mechanics, 32(8): 2539-2543,2560.

XIE Y M, LI T B, 2004. A preliminary study on the effect of blasting on rockburst[J]. The Chinese Journal of Geological Hazard and Control, 15(1): 64-67.

XIE H P, PARISEAU W G, 1993. Fractal characteristics and mechanism of rockburst [J]. Chinese Journal of Rock Mechanics and Engineering, (1): 28-37.

XU L S, 2004. Rockburst characteristics and prevention measures of Erlang Mountain Highway Tunnel[J]. China Civil Engineering Journal, 37(1): 61-64.

XU L S, WANG L S, LI Y L, 2002. Rockburst formation mechanism and criterion research [J]. Rock and Soil Mechanics, 23(3): 300-303.

XU L S, WANG L S, 2001. Research on Rockburst Formation Mechanism [J]. Journal of Chongqing University, 24(2): 115-117.

XU L S, WANG L S, 1999. Research on the occurrence and prediction of rockburst in Erlang Mountain Highway Tunnel [J]. Chinese Journal of Geotechnical Engineering, 21 (5): 569-572.

XU L S, WANG L S, LI T B, 1999. Summary of the research status of rockburst at home and abroad[J]. Journal of Yangtze River Scientific Research Institute, 16(4): 25-28,39.

XU J, ZHENG Y R, 2003. Elastoplastic stochastic finite element analysis and reliability calculation of tunnel surrounding rock[J]. Rock and Soil Mechanics, 24(1): 70-74.

XU X D, 2008. Preliminary Study on Mechanism of Rockburst Prevention in Tunnels [J]. Journal of the China Railway Society, 25 (10): 36-39.

XU Z M, HUANG R Q, FAN Z G, et al., 2004. Research progress on rockburst disasters in long tunnels[J]. Journal of Natural Disasters, 13(2): 16-24.

XU Z M, HUANG R Q, 2003. The relationship between rock blasting and blasting[J]. Chinese Journal of Rock Mechanics and Engineering, 22(3): 414-419.

YAN P, CHEN X R, SHAN Z G, et al., 2008. Research on rockburst prevention measures based on supershear stress control[J]. Rock and Soil Mechanics, 29(S1): 453-458.

ZHANG J M, 2003. Rockburst prediction and protection research [D]. Nanjing: Hohai University.

ZHANG J S, LU J Y, JIA Y R, 1991. Study on rock burst of diversion tunnel of Tianshengqiao II Hydropower Station[J]. Hydroelectric Power, (10): 34-37.

ZHANG J J, FU B J, 2008. Rock burst and its criterion and prevention[J]. Chinese Journal of Rock Mechanics and Engineering, 27(10): 2034-2042.

ZHANG Z Y, WANG S T, WANG L S, 1994. Principles of Engineering Geological Analysis [M].2nd ed. Beijing: Geological Publishing House.

ZHANG M T, 1987. Rockburst instability theory and numerical simulation calculation [J].

Chinese Journal of Rock Mechanics and Engineering, (3): 15-22.

ZHAO B J, 1995. Rockburst and its prevention[M]. Beijing: Coal Industry Press.

ZHAO S C, 2002. Resource Exploitation and Underground Engineering under Deep High Stress——Summary of the 175th Xiangshan Conference[J]. Advances in Earth Science, 17 (2): 295-298.

ZHOU R Z, 1995. Analysis of rockburst occurrence rules and fracture mechanics mechanism [J]. Chinese Journal of Geotechnical Engineering, 17 (6): 111-117.

ZHU F C, 2001. Study on the rockburst gestation process of hard rocks [D]. Changsha: Central South University.

ZHU L, LI H Z, WANG X M, et al., 2011. Influence of Explosive Material Performance Parameters on JPC Molding [J]. Journal of Sichuan Ordnance Industry, (3): 13-16.

ZUO Y J, LI X B, ZHAO G Y, 2005. Catastrophe model of cavern stratified buckling rockburst [J]. Journal of Central South University (Science and Technology), 36 (2): 311-316.